Collecting Oil Cans

W. Clark Miller & Sabra Sonewald

Schiffer Publishing Ltd®
4880 Lower Valley Road, Atglen, PA 19310 USA

Dedication

We would like to dedicate this book to our families and friends, who relentlessly search for items to add to our collection.

We lovingly embrace each piece and cherish the memories of its owners past and its discovery and transportation to us. Although we can't relay all the wonderful memories we have and the stories behind each and every item, we hope that you will enjoy seeing these treasures and feel a sense of history as you add similar items to your collection.

Disclaimer

The text and products pictured in this book are from the collection of the author of this book. It is not sponsored, endorsed, or otherwise affiliated with any of the companies whose products are represented herein. This book was derived from the author's independent research.

Library of Congress Card Number: 00-112274

Designed by John P. Cheek
Type set in Windsor BT/Souvenir Lt BT

ISBN: 0-7643-1336-3
Printed in China
1 2 3 4

Published by Schiffer Publishing Ltd.
4880 Lower Valley Road
Atglen, PA 19310
Phone: (610) 593-1777; Fax: (610) 593-2002
E-mail: Schifferbk@aol.com
Please visit our web site catalog at
www.schifferbooks.com or write for a free catalog.

We are always looking for authors to write books on new and related subjects. If you have an idea for a book, please contact us at the above address.

This book may be purchased from the publisher.
Please include $3.95 for shipping.

In Europe, Schiffer books are distributed by
Bushwood Books
6 Marksbury Ave.
Kew Gardens
Surrey TW9 4JF England
Phone: 44 (0)20-8392-8585;
Fax: 44 (0)20-8392-9876
E-mail: Bushwd@aol.com
Free postage in the UK. Europe: Air mail at cost.
Please try your bookstore first.

Contents

Acknowledgments

The authors would like to acknowledge the following people for their help over the years in expanding my collection: Mr. Scott Benjamin, Mr. Norm Griffin, Mr. J. Sam McIntyre, Jeff and Ellen Wood, Bill and Carla Sigworth, Tom and Char Whitlatch and my fellow collectors.

A special thanks to Mike Axel at Axel Photography.

Introduction

I have been interested in collecting cans as long as I can remember. I have always collected cans, maps and other small items when my Mom would stop at gas stations. My Dad worked at Lubrizol Corporation in Wickliffe, Ohio, developing additives for oil and gasoline and other types of fuels and lubricants. He would bring home product cans with his additives in them. I guess I liked the graphics and all the different brands even to the point of bugging my Mom to stop at stations we never dealt with to add cans to my collection. I still have about 100 of these cans in my collection.

I got hooked again after going to an antique show and picking up a couple of items. What started as an interest is now a full-blown passion, with a collection of 10,000 pieces of service station history! In this book I will share my vast can collection with you, including brands from Airline to Zeropruf, dating from the 1900s to the 1970s. I have tried to date each item as accurately as possible. In some cases exact dates can be determined. In other cases, dates can be approximated using the time period that the brand or logo was in use. I have included any important information in each photo's caption.

The prices quoted are for estimation only, as condition, rarity, and originality have a direct effect on prices of these cans, as with collectibles in general. The grading guide is just that. #9 condition to me may only be a #8 to someone else. I will leave the final decision to you, the collector. I hope this book is a valuable addition to your library. Happy collecting!

Oil Can Grading

Grade 10

This is the best condition a can can be in. This is most likely an unfilled display can with no surface scratches or rubbing from filling or shipping, and no discoloration from oil, chemicals, or sun fade. Both tops and bottoms are bright on paper or plastic cans. No damage to labels. All original seals will be intact. Of the 10,000 cans in my collection, only about 50 fall into this classification.

Grade 9

This can has scratches or rubbing that would have to be looked for, may have very minor sun fade.

Grade 8

Up to 10% of can, not normally on the face, may show light damage, including rubbing, scratches or sun fade.

Grade 7

This is the last grade in the excellent category. The can may show very heavy scratches, rubbing and fading or discoloration from oil or chemicals; top and bottom will show rust and discoloration.

Grade 6

More than 20%, but not over 40% of the surface will show scratches, rubbing, sun fade, and rust. The design will show and display well but may have pin holes in top and bottom.

Grade 5

Same condition as Grade 6 but will have dents over less than 40% of can. Pin hole rust in can face and body may be present.

Grade 4

More than 40% but not over 60% of the can will show scratches, rust, dents, sun fade and discoloration. The face of the can is still displayable.

Grade 3

60% to 80% of the can will have heavy damage with severe rust holes; up to 40% of the can will have damage to top and face.

Grade 2

Will show enough color and detail to be put in the right group of cans by brand. Major rust and/or damage.

Grade 1

This example is a filler until one in better shape comes along.

Grade 0

More than 90% damage throughout. This can may be kept if this is the only known example of this company or brand name.

Container Type Codes

Please note that not all of these codes were used in this book. Special thanks to Ed Love and Villa Publishing Syndicate for standardizing our hobby in the early years.

AL	Aluminum, thin walled cans, 1960s-1970s
CP	Composite paper / foil cans, 1945-1980s
CAR	Cardboard cans, WW II era, 1942-1945
CT	Cone top
PB	Plastic bottles
PC	Plastic cans
RS	Rib steel
SS	Straight sided steel, 1930-1970
VRT	Vertical rectangle steel

Known Can Companies

American Can Company
Boise-Cascade
Container Corporation of America
Continental Can Company
Crown Cork & Seal
Georgia-Pacific Corporation, Atlanta GA
Hardesty Can Company, Denver CO
National Can Company
Owens-Illinois
The Texas Company
Penn Plastic, Creighton PA
S.A.Y. Industries Inc., Leonminster MA
Spiro-PAC
St. Louis Can Company
Sonoco Products Company, Hartsville SC
Stone Container Company, Orlando FL
Weyerhaeuser Paper Company

Date Coding Tips

1947: Viscosity was listed only on can tops. No SAE designation.

1948: Series 2 classification by Caterpillar Tractor Company for diesel engines.

1952: ML, MM, MS, DG, DM and viscosity added to cans with SAE grade.

1953: 10W30 motor oils introduced.

1955: Series 3 classification by Caterpillar Tractor Company for diesel engines.

1963-64: Zip code added to cans.

1973: Bar codes (UPC) introduced.

1943-July 1, 1963: A one- or two-digit number between the city and state in the address is the postal zone used from 1943 to July 1, 1963. Example, the Shellzone Anti-Freeze one-quart can's title block reads New York 20 New York, San Francisco 6 California.

Note: The copyright date or patent numbers will also give an idea of the time period during which a can was produced.

Cans With Open Dates

Amoco

This coding starts about 1953. The code will appear on the back of the can towards the bottom. This example is from a 1961 vintage Permalube 1 quart.

262-24X-1q-FC-01/61

262: Brand name Permalube
24X: 24 cans to a case
1 qt: can size
FC: Not used for this purpose
01/61: Design first used January 1961

Atlantic

Any can listed as Atlantic Refining Company was produced before Spring 1966. Any can listing Atlantic Richfield was produced after Spring 1966.

BP

Any can listing Wilmington, Delaware dates 1969 and earlier; any listing Cleveland, Ohio is 1970 and on.

Deep Rock

Any Deep Rock can listing Kerr McGee is 1955 and on.

Esso

In July 1963 with the reformulation of all of its motor oil products, Humble Oil started using a code system on most of its product containers. The first letter is the division.

ESSO S
HUMBLE H
ENCO N

The number after the letter is the product.

UNIFLO 1
EXTRA 2
MOTOR OIL 3

The letter was added when the can design or formula was upgraded.

A, B, etc.

Example: Esso Extra S2B

Gulf

This code applies to all Gulf products from 1950-1985. Disregard all the letters in the code except for the last three or four. The first is the month code, and the second set is the year. The code is as listed below:

S P E C I A L T Y 0
1 2 3 4 5 6 7 8 9 0

Example from a Gulf Supreme one-quart can:

C4E-S S - A 0
1 1 6 0 = 11/60

This can design was introduced in November of 1960.

Mobiloil

1930-1955: Socony Vacuum Oil Company
1955-1966: Socony-Mobil Oil Company
1966-On: Mobil Oil Corporation

Pure Oil

Most Pure Oil cans from 1949 have a date code that appears on the back of the can, sometimes on the bottom or about one-half way up on the right side. This code is for a 1960 High HP Purelube one-quart can.

HPL-60-1

High HP Purelube, 1960, one quart. This design was introduced in 1960.

Spur

Cans dating 1/80-7/81 made by Container Corporation of America.

Sunoco

From 1937 on, Sunoco cans are dated on the front or the back towards the bottom.

Texaco

All Texaco cans from 1954 carry an open date on the reverse side.

Example: 9/54

Directory of Cans

Airline Oil Co., Chicago, IL

Airline Motor Oil, Airline Oil Co., Lubbock, TX one-quart can, 1936. $200-$600

American Oil Co., Chicago, IL

Amoco Motor Oil, 1920s (SS) two-gallon can. $125-$225

Amoco Motor Oil. Left: 1950 (SS). Right: 1940 (SS), one-quart can. $30-$50

Left: Counterclockwise, from top: AMO Anti Freeze, 1949 (SS). Amoco Permalube, 1940 (SS). Amoco Penn, 1940 (SS). All one-quart cans. $75-$250

Amoco Penn, 1951 (SS), one-quart can. $25-$40

Amoco Super Permalube, 1959 (SS), half-quart can. $50-$75

Amoco Permalube, Left: 1954 (SS). Right: 1956 (SS), one-quart can. $25-$40

Amoco Motor Oil. Left: 1954 (SS). Right: 1956 (SS), one-quart can. $25-$40

Amoco Penn, Left: 1954 (SS). Right: 1956 (SS), one-quart can. $25-$40

Super Permalube, Left: 12/53 (SS). Right: 4/54 (SS), one-quart can. $30-$60

Super Permalube, 11/56 (SS), one-quart can. $25-$45

Left: Permalube Subzero, 9/52 (SS), one-quart can. $30-$60. Right: Permalube All Weather, 6/53 (SS), one-quart can. $30-$60

Standard Super Permalube, five-gallon can, 11/56. $25-$40

Permalube, Top: 2/55 (SS). Left: 1950 (SS). Right: 1/53 (SS). Not shown 8/59 (SS), one-quart cans. $25-$45 each

ISO-VIS, Top: 1940 (SS), Left: 1945 (SS). Right: 1946 (SS), one-quart cans. $30-$60

Polarine. Top: 1940 (SS), Left: 1945 (SS). Right: 7/52 (SS). Not shown, 1948 (SS). One-quart cans, $30-$60

Left: Stanolube, 1948 (SS), $50-$75. Right: Stanolube HD-M, 7/52 wide seam (SS), one-quart can, $40-$70.

Standard ISO-VIS, 7/52, five-gallon can (SS). $25-$50

Stanolube HD-M. Left: 7/52 (SS) Folded seam. Right: 7/52 (Silver can) (SS). One-quart cans, $30-$60.

American LDO, 8/64 (CP). Not shown: 1962 (SS). One-quart can, $100-$175. Rare.

Permalube. Top: 9/66 (CP). Left: 1/61 (SS). Right: 11/61 (CP). One-quart cans, $25-$35.

Super Permalube LDO. Left: 9/66 5W30. Right: 9/66 10W 20W 30. Both (CP), one-quart cans, $40-$60.

Amolube. Top: 9/66 (CP). Left: 4/61 (SS). Right: 11/61 (CP). One-quart cans, $20-$35.

Super Permalube. Top: 9/66 (CP). Left: 1/61 (SS). Right: 8/64 (CP). One-quart cans, $30-$45.

Artex. Left: 1/61 (SS). Right: 1/69 (CP). One-quart cans, $50-$75. Rare

American S-1, 1/61 (SS), one-quart can. $25-$40

American Truck & Bus, 11/61 (CP), one-quart can. $25-$40

American S-3. Left: 1/61 (SS). Right: 8/64 (CP). One-quart cans, $25-$40.

American HD-M. Top: 8/64 (CP). Left: 1/61 (SS). Right: 11/61 (CP). One-quart cans. $25-$40

Left to right: Amoco 100, 8/67 (CP); Amoco 200, 8/67 (CP); Amoco 300, 5/66 (CP); Amoco MV, 1/69 (CP). One-quart cans, $25-$35 each.

Amoco ATF A. Top: 8/64 (CP). Left: 7/58 (SS). Right: 1/61 (SS). One-quart cans, $30-$50.

Standard Outboard, resealable quart, 3/60 (SS). $30-$40

Left: Amoco ATF F, 9/67 (CP). Right: Amoco ATF Dexron, 9/67 (CP). One-quart cans. $20-$30.

Standard/American Outboard. Left: 9/51 (SS) one-quart can. Right: 1/61 (SS), one-quart can. $25-$35

Amoco Outboard Oil. Left to right: 1950 (SS) one-quart can, 1954 (SS) one-quart can, 1956 (SS) one-quart can. $35-$45 each.

American Outboard. Left: 10/64 (CP). Right: 7/67 (CP). One-quart cans, $35-$35.

Standard Aviation, 1955 (SS), one-quart can. $100-$300

American Outboard, plastic quart, 3/69. $20-$30

Amoco Super Anti-Freeze, 1949 (SS), one-quart can. $75-$150

American Snowmobile, 9/68 (CP), one-quart can. $40-$60

Standard Super Anti-Freeze. Left: 1950 (SS). Right: 1958 (SS). One-quart cans, $30-$60.

Artex two-gallon can, 1/61 (SS). $25-$40

Amoco Artex, 1938 (SS), two-gallon can. $35-$45

Stanolind two-gallon cans. Left: 1947 (SS). Right: 1956 (SS). $25-$40

Pan-Am Motor Oil, 1946 (SS), one-quart can. $75-$150

Pan Panoco, 1948 (SS), one-quart can.
$75-$150

Utoco HD-M, 1955 (SS), one-quart can.
$75-$225

Pan-Am Permalube, 1953 (SS), five-quart can.
$50-$200. Also sold in quart.

Utoco Permalube, 1945 (SS), one-quart can. $150-$250

Pan-Am Grease.1940s-1950s (SS) 1-lb. cans. $35-$50 each.

Utoco Permalube, 1955 (SS), one-quart can.
$75-$225

Eureka Harness Oil, Standard Oil of Indiana, 1929 (S), five-gallon can. $75-$150

Apex Oil Co.

Film X, Apex Oil Co., 1946 (SS), one-quart can. $175-$375

Atlantic Richfield Oil Co., New York, NY

Atlantic Motor Oil for Fords, 1922 (SS) one-gallon can. $200-$400

Atlantic Aviation, 1940 (SS), one-quart can. Aviation is the brand name of the oil and was not sold for use in aircraft. $125-$275

Atlantic Aviation. Left to right: 1946 one-quart can (SS), 1949 one-quart can (RS), 1952 one-quart can (SS). $40-$60

Atlantic Aviation. Left to right: 1958 one-quart can (SS), 1962 one-quart can (CP), 1964 one-quart can (CP). $25-$35

Atlantic Motor Oil. Left to right: 1952 one-quart can (SS), 1949 one-quart can (RS), 1950 one-quart can (SS). $25-$35

Atlantic Aviation. Left to right: 1955 one-quart can (SS),1966 one-quart can (CP), 1967 one-quart can (CP). $25-$35

Atlantic Motor Oil. Left to right: 1961 one-quart can (SS), 1964 one-quart can (CP), 1966 one-quart can (CP). $25-$35

Atlantic Motor Oil. Left to right: 1949 one-quart can (SS), variation with state of Florida on side. 1955 one-quart can (SS). 1958 one-quart can (SS). $25-$35

Atlantic Premium. Top: 1961 one-quart can (CP). Left: 1958 one-quart can (SS). Right: 1960 one-quart can (SS). $25-$35

Atlantic Imperial. Left: 1963 (CP). Right: 1966 (CP). $20-$30

Atlantic Outboard, 1957 one-quart can (SS). $30-$60

Atlantic Outboard. Left: 1963 (PB). Right: 1966 (PB). $15-$25

Atlantic ATF. Top: 1960 one-quart can (SS). Left: 1963 one-quart can (CP). Right: 1966 one-quart can (CP). $20-$30

Atlantic Capitol, 1952 (SS) two-gallon can. $25-$35

Atlantic Capitol. 1961 (SS) two-gallon can. $25-$35

Left: Richlube Motor Oil, 1961 one-quart can (SS). Right: Richfield-Penn, 1966 one-quart can (CP). $75-$125

Atlantic. Left to right: 1960 (SS) 4 oz., 1966 (SS) 4 oz., 1966 (SS) 4 oz. $30-$40

Left: Richlube Premium, 1954 one-quart can (SS). Right: Richlube, 1966 one-quart can (CP). $40-$75

Richfield Motor Oil. These are from Richfield Eastern Division New York, New York. Top: 1964 one-quart can (CP). Left: Richlube 10W30, 1961 one-quart can (SS). Right: Richlube Super HD, 1961 one-quart can (SS). $40-$60

Left: Richlube Premium, 1965 one-quart can (CP). Right: Richfield-Penn, 1961 one-quart can (SS). $75-$125

Left: Richfield Penn, 1956 one-quart can (SS). Right: Richlube Extra Duty, 1958 one-quart can (AL). $50-$75

Richzone Anti-Freeze, 1961 one-gallon can (SS). Richfield Oil Co., New York, New York. $25-$35

Atlas Supply Co., Springfield, NJ

Atlas was a company established in 1931 to supply all Standard Oil divisions with tires, batteries and other accessory items.

Atlas Coldex Anti-Freeze. Left: 1955 one-quart can (SS). Right: 1960 one-quart can (SS). $25-$35

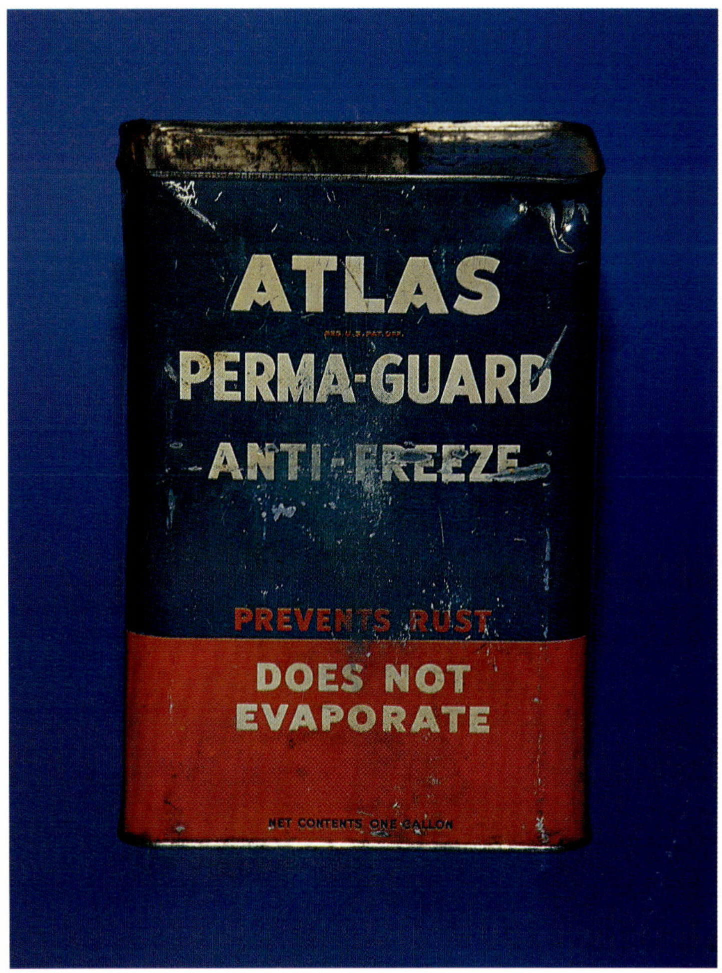

Atlas Permaguard, 1939 1-gallon can (SS). $25-$35

Atlas Permaguard, Left: 1952 1-gallon can (SS). Right: 1954 1-gallon can (SS). $25-$30

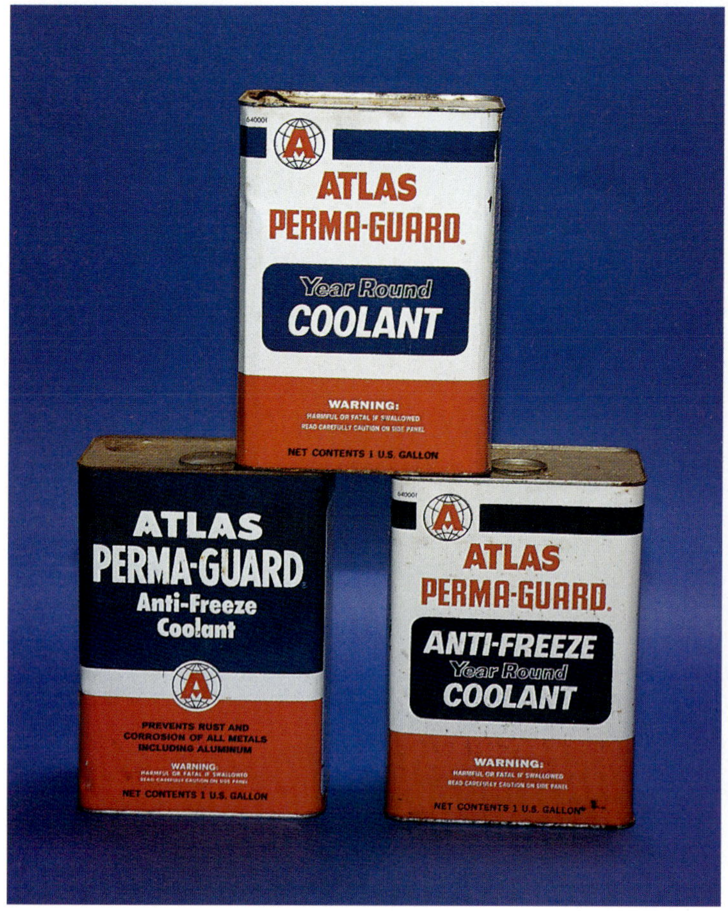

Atlas Permaguard. Top and right: 1971 one-gallon can (SS). Front and back shown. Left: 1969 1-gallon can (SS). $15-$25

Atlas Frostguard. Top: 1954 1-gallon can (SS). $15-$25. Atlas Permaguard, Left: 1956 1-gallon can (SS). Right: 1957 1-gallon (SS). $15-$30

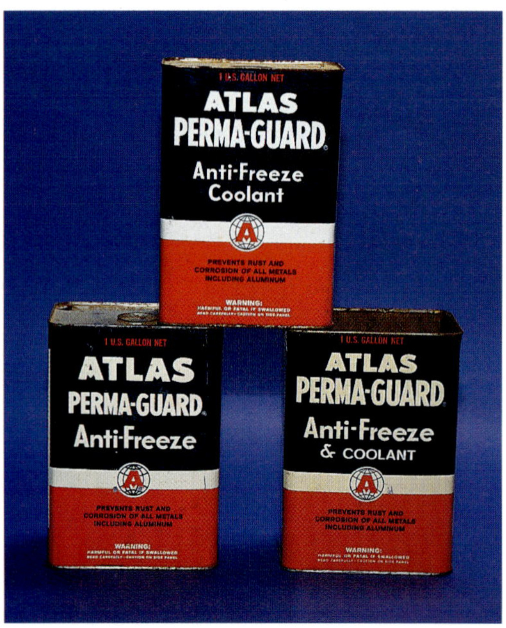

Atlas Permaguard. Top: 1968 1-gallon can (SS). Left: 1963 1-gallon can (SS). Right: 1966 1-gallon can (SS). $15-$25

A.W. Harris Oil Co., Providence, RI

Harris Motor Oil, A.W. Harris Oil Co., Providence, Rhode Island. 1940s (SS) one-quart can. $225-$325

Barnsdall Oil Co.

Barnsdall, No company location, 1946 one-quart can (SS). $50-$75

Bodie Hoover Petroleum Co., Lemont, IL

Motor Seal, Bodie, Hoover Petroleum Co., Lemont IL. 1961 one-quart can (SS). $75-$150

Benz Oil Co., Milwaukee, WI

Tulane Motor Oil, Benz Oil Co., Milwaukee WI. 1958 one-quart can (SS). $75-$150

Cato Oil & Grease Co., Oklahoma City, OK

Penntroleum, Cato Oil & Grease Co., Oklahoma City, Oklahoma. 1950 one-quart can (SS). $75-$250

Champlin Oil Co., Enid, OK

Champlin Motor Oil, Left: 1948 (SS). Right: 1956 (SS). One-quart cans, $25-$35

Champlin HI-VI, 3 variations 1956-1961, all (SS). One-quart cans, $30-$40.

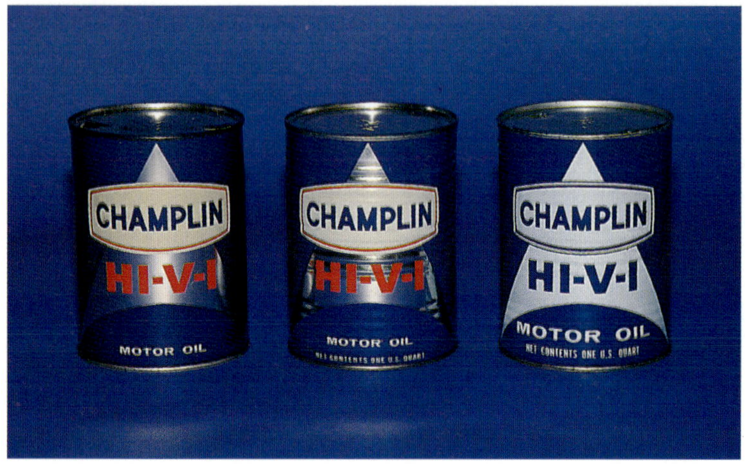

Champlin HI-VI. Left: 1961 (SS). Middle: 1964 (RS). Right: 1969 (RS). One-quart cans, $25-$35.

Champlin Trans Season. Left: 5W30 1963 (RS). Right: 10W30 1963 (RS). One-quart cans. $25-$35

Champlin Delux. Left: 10W30, 1960 (SS). Middle: Motor Oil, 1960 (SS). Right: Improved, 1964 (RS). One-quart cans, $25-$35.

Champlin CMO, Left: 1964 (RS), Right: 1969 (RS). One-quart cans, $25-$35.

Champlin S-3. Left: 1964 (RS). Right: 1969 (RS). One-quart cans, $25-$35.

Champlin LPG. Left: 1956 (SS). Middle: 1964 (RS). Right: 1969 (RS). One-quart cans, $30-$40.

Left to right: Champlin Aircraft, 1956 (SS) one-quart can, $100-$200. 1966 Champlin Snowmobile (RS), one-quart can, $50-$75. Champlin Anti-Freeze, 1946 (SS) one-quart can, $50-$75.

Champlin ATF A, 1964 (RS), one-quart can. $25-$35

Champlin Outboard, 1961 (SS) one-quart can. $25-$35

C.H. Clark Oil Co., Cleveland OH

Sea Gull Motor Oil, C.H. Clark Oil Co., Cleveland Ohio. 1945 (SS) one-quart can. $350-$450

Zerolene, Zero Cold Test. 1920s (SS) five-gallon can. $175-$375

Chevron Oil Co., (Standard Oil Co. California)

Western Division, San Francisco, CA; Denver, CO
Eastern Division, Perth Amboy, NJ
Standard Oil of Kentucky Division, Louisville, KY

RPM Thermo-Charged, 1938 (SS), one-quart can. $50-$75

RPM Motor Oil, 1946 (SS), one-quart can. $50-$75

RPM 10-40 Special. Left: 1956 (SS). Right: 1957 (SS). One-quart cans, $25-$35 each.

RPM 10-30 Special. Left: 1956 (SS). Right: 1957 (SS). One-quart cans, $25-$35 each.

RPM Motor Oil. Top: 1951 (SS). Right: 1947 (SS). Left: 1948 (SS). One-quart cans, $30-$40 each.

RPM Heavy Duty. Top: 1947 (SS). Right: 1950 (SS). Left: RPM Motor Oil, 1949 (S). One-quart cans, $30-$40 each.

RPM Heavy Duty. Left: 1956 (SS). Right: 1958 (SS). One-quart cans, $25-$35 each.

RPM Custom. Left: 1959 (SS). Right: 1967 (CP). One-quart cans, $30-$50 each.

RPM Supreme. Left: 1956 (SS). Right: 1963 (SS). One-quart cans, $25-$35 each.

RPM Special. Left: 1960 (SS). Right: 1961 (SS). One-quart cans, $25-$35 each.

RPM Supreme. Left: 1959 (SS). Right: 1965 (CP). One-quart cans, $25-$35 each.

RPM Special. Left: 1965 (CP). Right: 1969 (CP). One-quart cans, $20-$30 each.

RPM Special. Top: 1956 (SS). Left: 1959 (SS). Right: 1963 (SS). One-quart cans, $25-$35 each.

Zerolene. Left: 1954 (SS). Right: 1956 (SS). One-quart cans, $30-$50 each.

Zerolene. Left: 1961 (SS). Right: 1965 (CP). One-quart cans, $25-$35 each.

Zerolene. Left: 1966 (CP). Right: 1969 (CP). One-quart cans, $25-$35 each.

Standard Penn. Left: 1935 (SS). Right: 1967 (CP). One-quart cans, $25-$125.

Standard Penn. Top: 1950 (SS). Left: 1955 (SS). Right: 1958 (SS). One-quart cans, $30-$50 each.

S.O. Penn. Left: 1961 (SS). Right: 1969 (CP). Sold only through Chevron stations. One-quart cans, $25-$35.

Standard Penn, 1969 (CP). Sold only through Standard branded stations. One-quart can, $25-$45.

Calol. Left: 1961 (SS). Right: 1969 (CP). Sold only at Standard Oil of KY stations. One-quart cans, $30-$50.

RPM Delo Special. 1955 (SS), one-quart can. $25-$35. RPM Supercharged-1, 1955 (SS), one-quart can. $40-$60

RPM Delo. Left: 1947 (SS). Right: 1948 (SS). One-quart cans, $40-$40 each.

Super RPM Delo. Top: 1969 (CP). Left: 1966 (CP). Right: 1961 (SS). One-quart cans, $25-$35 each.

RPM Delo Super, 1954 (SS), two variations. One-quart cans, $35-$45.

RPM Multi Service. Left: 1963 (CP), one-quart can, $25-$35. Right: Super 3 Oil, 1965 (CP), one-quart can, $25-$40.

RPM Sub Zero, 1951 (SS), one-quart can. $75-$150

RPM Aviation, 1963 (SS), one-quart can. $25-$35 each.

RPM Aviation, 1955 (SS), one-quart can. $75-$175

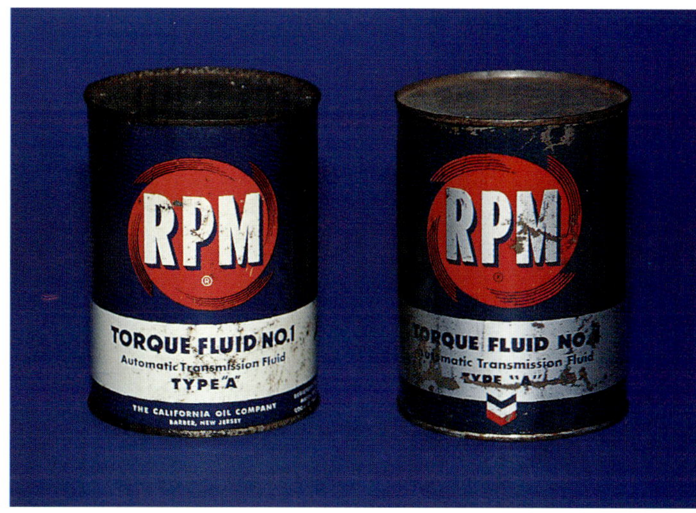

RPM Torque Fluid. Left: 1951 (SS). Right: 1956 (SS). One-quart cans, $25-$35.

RPM Aviation, 1956 (SS), one-quart can. $35-$45 each.

RPM ATF. Left: 1961 (SS). Right: 1969 (CP). One-quart cans, $25-$35.

Left: Caravan two-gallon can, Standard Oil California, Eastern Division, 1958 (SS), $30-$50. Right: Calso two-gallon can, S.O California, Eastern Division, 1961 (SS), $30-$50.

RPM Outboard, 1960 (SS), one-quart can. $45-$65

RPM Outboard, Left to right: 1949 (SS), 1953 (SS), 1955 (SS). $25-$35

RPM Outboard. Left: 1958 (SS) one-quart can. Right: 1960 (SS) one-quart can. $25-$35

Cities Service Oil Co., New York, NY

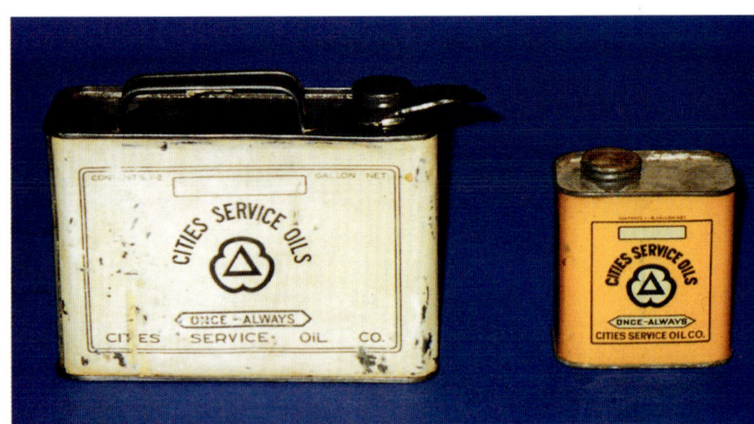

Cities Service Motor Oil. Left: 1920s (SS) one-gallon can. Right: 1920s (SS) one-pint can. $100-$200 each.

Top: Cities Service Motor Oil, 1934 (SS). Left: Cities Service Koolmotor, 1936 (SS). Right: Cities Service Koolmotor, 1934 (SS), one-quart cans. $30-$60

Cities Service Motor Oil. Left: 1936 (SS). Right: 1950 (SS). One-quart cans, $30-$50.

Cities Service Trojan. Top: 1953 (SS). Left: 1945 (SS). Right: 1947 (SS). One-quart cans, $30-$60.

Cities Service Trojan. Top: 1963 (CP). Left: 1961 (SS). Right: 1965 (CP). One-quart cans, $20-$35.

Cities Service Koolmotor. Left to right: 1943, 1948, 1954, 1955, all (SS). One-quart cans, $30-$50 each.

Cities Service HD Koolmotor. Left: 1957 (SS). Right 1960 (SS). One-quart cans, $25-$35.

Cities Service HD Koolmotor. Left: 1962 (SS). Right: 1964 (CP). One-quart cans, $25-$40.

Cities Service 5D. Left: 1954 (SS). Right: 1956 (SS). One-quart cans, $40-$150.

Cities Service 5D, all 1956 (SS), one-quart cans. $25-$40

Cities Service 5D. Top: 1961 (SS). Left: 1957 (SS) "New". Right: 1958 (SS). One-quart cans, $25-$45.

Cities Service C Series, Left: 1956 (SS). Right: 1957 (SS). All C Series cans are stamped on top 300 for S-1 and 500 for S-3 grade. One-quart cans, $25-$35.

Cities Service C Series. Left: 1961 (SS). Right 1965 (CP). One-quart cans, $25-$40.

Cities Service Blue Club, 1951 (SS), two-gallon can. $35-$65

Cities Service Blue Club, 1955 (SS), one-quart can. Also sold in two-gallon cans. $100-$250. Rare.

Cities Service Blue Club. 1956 (SS) two-gallon can. $25-$35

Cities Service Blue Club, two-gallon (SS), 1957. $25-$40

Cities Service Blue Club, two-gallon. Left: 1963 (SS). Right: 1961 (SS). $25-$40

Cities Service Outboard, Left to right: 1954 (SS) one-quart can, 1955 (SS) one-quart can, 1950 (SS) one-quart can. $35-$125

Cities Service Outboard. Cities Service sold two grades of outboard oil at this time. Left: 1963 (SS) one-quart can. Right: 1963 (SS) one-quart can. $25-$35

Left: Cities Service Outboard, 1955 (SS), $125-$325. Right: 1965 (CP), $50-$125.

Cities Service ATF A. Left: 1956 (SS), Right: 1953 (SS). One-quart cans. $30-$60

Cities Service ATF A. Left: 1957 (SS). Right: 1963 (SS). One-quart cans, $30-$40.

Left: Cities Service Aero Oil, 1948 (SS) one-quart can, $200-$400. Right: Cities Service Koldpruf, 1950s (SS) one-quart can, $75-$150.

Cities Service Koldpruf. Left: 1939 (SS), Right: 1947 (SS). One-quart cans, $125-$225.

Cities Service Anti-Freeze, 1948 (SS), one-quart can. $300-$600

Cities Service Anti-Freeze, 1953 (SS), one-gallon can. $30-$40

Cities Service Koldpruf. Left: 1956 (SS). Right: 1957 (SS). One-quart cans, $50-$100.

Top: Cities Service Anti-Freeze, 1953 (SS). Left: Cities Service Koldpruf, 1953 (SS). Right: Cities Service Koldpruf, 1955 (SS). One-quart cans, $75-$150.

Cities Service Koldpruf. Left: 1963 (SS). Right: 1961 (SS). One-quart cans. $30-$50

Cities Service Koldpruf. 1955 (SS) one-gallon can. $25-$35

Citgo Extra Range. Left: 1971 (CP). Right: 1966 (SS). One-quart cans, $25-$35.

Citgo Premium Plus, 1966 (CP), one-gallon can. Also sold in quarts. $25-$35

Citgo Milemaster, 1966 (SS), one-quart can. $25-$35

Citgo Premium, 1966 (CP), one-quart can. $25-$35

Citgo Regular, 1966 (CP), one-quart can. $25-$35

Continental Oil Co., Houston, TX

Conoco Germ Processed, 1930-1940, one-quart cans, (SS). There are nine variations of this can. $25-$75

Conoco Nth. Left: 1940 (SS), Right: 1944 (SS). One-quart cans, $25-$35.

Conoco Nth. Left: 1945 (SS). Right: 1948 (SS). One-quart cans, $25-$35.

Conoco All Season Super. Top: 1964 (CP). Left: 1959 (SS). Right: 1962 (SS). One-quart cans, $25-$35.

Left: Conoco Super Heavy Duty, 1954 (SS), $25-$35. Right: Conoco All Season Super, 1955 (S), $25-$35.

Conoco All Season Super. Left: 1966 (CP). Right 1969 (CP). One-quart cans, $30-$50.

Conoco Super. Left: 1953 (SS). Right: 1950 (SS). One-quart cans, $25-$45.

Conoco Super. Top: 1955 (SS). Left: 1954 (SS). Right: 1956 (SS). One-quart cans. $35-$50.

Conoco Super, 1966-1971, five variations, all (CP). $15-$25

Conoco Super. Top: 1964 (SS). Left: 1959 (SS). Right: 1964 (CP). One-quart cans, $25-$35.

Conoco Motorine. Left: 1936 (SS), $50-$75. Right: 1944 (SS), $35-$45.

Conoco Motorine. Left: 1950 (SS). Right: 1960 (SS). One-quart cans, $35-$45.

Conoco Motorine HD, 1960 (SS), $25-$40.

Conoco HD, 1955 (SS). $30-$50

Conoco Par, 1956 (SS). $50-$75

Conoco Par. Top: 1964 (CP). Left: 1960 (SS). Right: 1963 (SS). One-quart cans, $20-$40.

Conoco Par, 1966-1971. Three variations, all (CP). $15-$25

Conoco Motor Oil, 1950s (SS). These cans were used for all commercial oils, with product stamped on lid. $25-$35

Conoco Motor Oil. Left: 1966 S-3 (SS). Right: 1969 S-3 (SS). One-quart cans. $25-$35

Conoco Motor Oil. Top: 1970 Tracon (SS). Left: 1970 Tracon Supreme (SS). Right: 1971 S-3 (SS). $25-$35 each.

Conoco Outboard. Left: 1960 (SS). Right: 1964 (SS). One-quart cans, $35-$50.

Conoco Outboard. Left: 1953 one-quart can (SS). Right: 1955 1-pint can (SS). $35-$55

Conoco Outboard. Left: 1957 (SS) one-quart can. Right 1960 (SS) one-quart can. $25-$35

Conoco Scat Oil. Left: 1966 (SS), one-quart can, $75-$100. Right: 1972 (SS), one-quart can, $20-$30.

Conoco H-D, 1952 (SS), one-quart can. $25-$40

Conoco Aero S, 1960 (SS), one-quart can. $75-$150

Conoco Aero S. Top: 1970 (CP), $25-$45. Left: 1966 (SS), one-quart can, $25-$50. Right: 1969 (SS), one-quart can, $25-$50.

Conoco ATF Type A, 1955 (SS), one-quart can. $25-$50

Conoco ATF A. Top: 1964 (CP). Left: 1961 (SS). Right: 1963 (SS). One-quart cans, $25-$45.

Conoco ATF A, 1966 (CP), one-quart can. $25-$35

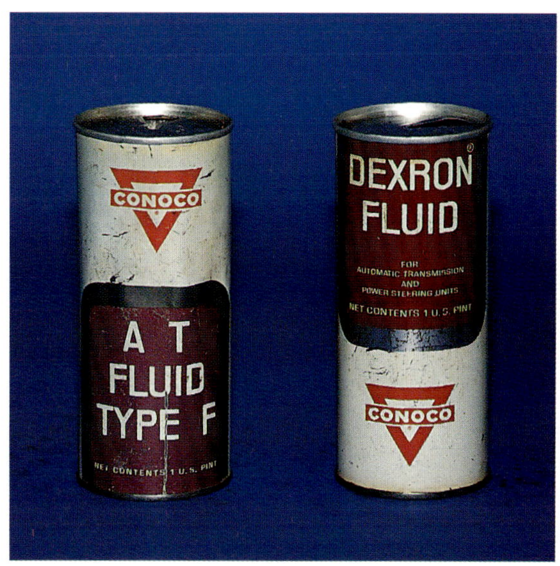

Conoco ATF, pint cans, 1969 (SS). Also sold in quarts. $25-$35

Conoco Demand, 1940s (SS) two-gallon can. $40-$60

Left: Conoco Anti-Freeze, 1963 (SS), $50-$75. Right: Conoco Anti-Freeze, 1964 (S), $25-$35.

Conoco Anti-Freeze. Left: 1958 (SS). Right: 1955 (SS). One-quart cans, $30-$50 each.

Conoco Germ Process, 1936 (SS), 4 oz. $75-$125

Conoco Anti-Freeze, 1955 (SS), 1-gallon can. Also sold in quarts. $30-$50

Conoco, left to right: 1954 (SS) 4 oz., 1956 (SS) 4 oz., 1956 (SS) 4 oz. $25-$40 each.

Left: Conoco Household Oil, 1960 (SS), 4 oz., $20-$30. Right: Conoco Lighter Fluid, 1960 (SS), 4 oz., $30-$40.

Conoco, left to right: 1966 (SS) 4 oz., 1969 (SS) 4 oz.,1969 (SS) 4 oz. $15-$25 each.

Consumers COOP Association, Kansas City, MO

COOP Heavy Duty, Consumers COOP Association, Kansas City, Missouri. 1958 one-quart can (SS). $25-$35

United Cooperatives, Alliance, OH

Unico Motor Oil, 1948 five-gallon can (SS). $75-$125

Unico Bureau Premium. Left: 1959 one-quart can (SS). Right: 1965 one-quart can (SS). $40-$60 each.

Unico Series 3. Left: 1969 one-quart can (CP). Right: 1973 one-quart can (CP). $20-$30 each.

1959 five-gallon can (SS). $40-$60

Unico Outboard, 1963 one-quart can (SS). $25-$40

Unico Motor Oils. Left to right: 1973 one-quart can (CP). 1971 one-quart can (CP). 1971 one-quart can (CP). 1971 one-quart can (CP). 1974 one-quart can (CP). $15-25

Unico ATF. Left: 1971 one-quart can (Dextron) (CP). Right: 1971 one-quart can (Type F) (CP). $15-$25

Unico Methanol Anti-Freeze. 1958 (SS) one-gallon can. $20-$35

Unico Anti-Freeze. Top: 1971 one-quart can (SS). Left: 1964 one-quart can (SS). Right: 1960 one-quart can (SS). $30-$60

Unico Anti-Freeze, 1967 (SS) one-gallon can. $25-$35

Unico Permanent Anti-Freeze and Unico Methanol Anti-Freeze. 1950s-1960s (SS) one-gallon cans. $20-$35

Unico Anti-Freeze. 1960s-early 1970s (SS) one-gallon can. $20-$30

Unico Grease, 10 lb., can, 1948 (SS).
$30-$60

Left: Unico Gear Oil, 1971 (SS) two-gallon can. Right: Unico Regular Gear Oil, 1969 (SS) two-gallon can. $25-$35 each.

Unico Motor Oil, 1963 (SS) two-gallon can.
$25-$35

Left: Unico R&O Hydraulic Oil, 1971 (SS) two-gallon can. Right: Unico Special Motor Oil, 1971 (SS) two-gallon can. $25-$35 each.

Unico Household Oil. 1947 (SS) 8-oz. can. $40-$60

John Deere, Moline, Illinois

John Deere Torq Gard, John Deere, Moline, Illinois. 1975 one-quart can (CP). $10-$20

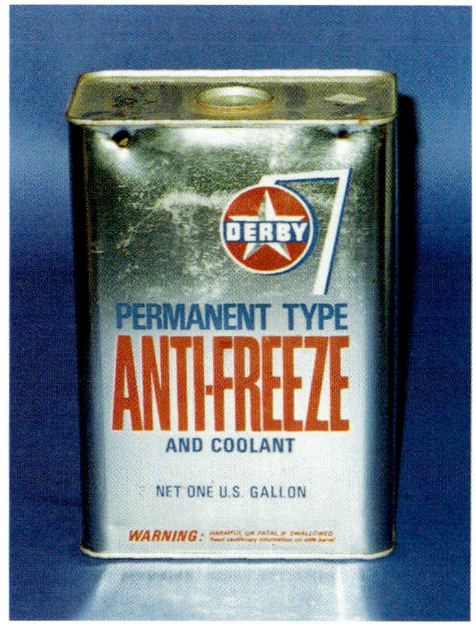

Derby Anti-Freeze. 1969 (SS) one-gallon can. $20-$30

Derby Oil Co., Wichita, Kansas

Derby Oil Co., Wichita, Kansas. Left to right: 1961 (SS) one-quart can, 1961 (SS) one-quart can, 1961 (SS) one-quart can, 1969 (CP) one-quart can. $25-$35 each.

Derby Anti-Freeze. Left: 1966 (SS) one-quart can. Right: 1969 (SS) one-quart can. $25-$35

Derby Outboard, 1961 (SS) one-quart can. $25-$35

Dryer, Clark, Dryer Oil Co., Oklahoma City, Oklahoma

Hercules, Dryer, Clark, Dryer Oil Co., Oklahoma City, Oklahoma. 1958 one-quart can (SS). $25-$35

DX Sunray Oil Co., Tulsa, OK

DX Diamond 760. Left: 1938 (SS). Right: 1941 (SS). One-quart cans. $30-$50

Left: DX Diamond, 1945 (SS). Right: DX Power, 1945 (SS). One-quart cans, $30-$50.

Left: D-X Motor Oil, 1955 (SS), $25-$45. Right: D-X DHD, 1955 (SS), $25-$35.

D-X Heavy Duty. Left: 1955 (SS), Right 1953 (SS). One-quart cans, $25-$35.

DX Heavy Duty. Top: 1964 (CP). Left: 1963 (SS). Right: 1960 (RS). One-quart cans, $25-$35.

D-X Special. Left: 1954 (SS). Right: 1956 (SS). One-quart cans, $25-$45.

DX Boron Extra Heavy Duty. Left: 1961 (AL). Right: 1964 (CP). One-quart cans, $25-$35.

DX Boron Ultra Special. Top Left: 1964 (CP).Top Right: 1970 (CP). Bottom Left: 1963 (SS). Bottom Right: 1961 (AL). One-quart cans, $25-$35 each.

DX Heavy Duty. Left: 1960 (SS). Right: 1963 (AL). One-quart cans, $25-$35.

Left: DX Extra Heavy Duty, 1957 (SS), one-quart can, $30-$50. Right: DX Boron Extra Heavy Duty, 1963 (SS), one-quart can, $25-$35.

DX Diamond. Left: 1969 (CP) one-quart can. Right: 1970 (CP) one-quart can, $15-$20

DX Power. Top: 1969 (CP). Left: 1963 (SS). Right: 1961 (AL). One-quart cans, $25-$35 each.

DX Distance Supreme. Left: 1964 (CP). Right: 1969 (CP). One-quart cans, $25-$50 each.

DX Super Sport, 1969 (CP) one-quart can. $25-$50

DX Power, 1961 (SS) two-gallon can. $30-$40

Right: DX Kleenring S-3. Top: 1964 (CP). Left: 1963 (SS). Right: 1961 (AL). One-quart cans, $25-$35 each

Left to right: DX Marine Oil, 1955 (SS) one-quart can, $40-$60. DX Outboard, 1955 (SS) one-quart can, $25-35. DX Outboard, 1957 (SS) one-quart can, $25-$35.

D-X Marine Oil, 1955 (SS) one-quart can. $150-$350

DX Anti-Freeze, 1969 (SS), one-gallon can. $25-$35

DX Outboard, 1960 plastic can, one quart. $50-$75

DX ATF Type A, 1955 (SS), one-quart can. $25-$40

DX ATF Type A. Left: 1963 (SS). Right: 1961 (AL). One-quart cans, $25-$45.

Top: DX ATF Type F, 1969 (CP), $25-$35. Left: 1969 (CP) Dextron. Right: 1964 (CP) Type A. One-quart cans, $25-$35.

Esso Standard (Standard Oil Co. NJ)

Pennsylvania Lubricating Co., Pittsburgh, PA

Uniflo Winter, 1934 (SS), one-quart can. $100-$200

Esso Motor Oil. Left: 1935 (SS). Right: 1939 (SS). One-quart cans, $30-$35 each.

Esso Motor Oil. Top, left to right: (AL), (CP). Bottom, left to right: (SS), (AL), (AL), (AL). One-quart cans, 1949-1965. $25-$35 each.

Esso Motor Oil. Top, left to right: (SS), (SS). Bottom, left to right: (CP), (SS), (CP). 1962-1966, one-quart cans. $20-$30 each.

Esso Extra, 1949, one-quart cans (SS), two variations. $35-$55

Esso Extra. Left: 1950 (SS). Right: 1955 (SS). One-quart cans, $25-$35.

Esso Extra, 1960 one-quart cans (both SS), two variations. $25-$35

Esso Extra, 1962, one-quart cans, two variations, (SS), (AL). $25-$35

Esso Extra. Top: 1963 (SS). Left: 1962 (CP). Right: 1963 variation. One-quart cans, $25-$35.

Esso Extra, 1958 (AL), one-quart cans, two variations. $25-$35

Esso Extra. Left: 1964 (CP). Right: 1965 (SS). One-quart cans, $25-$35.

Esso Extra. Left: 1960 (AL). Right: 1961 (SS). One-quart cans, $25-$35.

Esso Extra. Left: 1966 (SS). Right: 1967 (CP). One-quart cans. $25-$35

Esso Uniflo, 1954-1960. Top: (SS), (SS). Bottom: (SS), (AL). One-quart cans, $25-$35.

Essolube. Top: 1965 (SS). Left: 1947 (SS). Right: 1963 (SS). One-quart cans, $25-$75.

Esso Uniflo. Top: 1965 (SS). Left: 1963 (SS). Right: 1964 (SS). One-quart cans, $25-$35.

Essolube. Left: 1966 (SS). Right: 1967 (CP). One-quart cans, $25-$35.

Essolube, 1936 (SS), one-quart can. Has top variations. $30-$50

Essolube HD. Left: 1945 (SS). Right: 1953 (SS). One-quart cans, $35-$65.

Essolube HDX. Top: 1958 (AL). Left: 1960 (AL). Right: 1961 (SS). One-quart cans, $30-$40.

Essolube HDX. Left: 1964 (SS). Right: 1962 (SS). One-quart cans, $25-$35.

Essolube HDX. Left: 1966 (SS). Right: 1967 (CP). One-quart cans, $25-$35.

Actol Motor Oil. Left: 1960 (SS). Right: 1953 (SS). One-quart cans, rare, $50-$100.

Actol. Left: 1956 (SS) two-gallon can. Right 1961 (SS) two-gallon can. $35-$45 each.

Esso Aviation. Left: 1961 (SS). Right: 1965 (SS). One-quart cans, $25-$35.

Esso Aviation. Left: 1938 (SS). Right: 1945 (SS). One-quart cans. $200-$400 each

Esso Aviation, Left: 1954 (SS), Right 1959 (SS). One-quart can. $25-$35

Esso ATF. Top: 1960 (SS). Left: 1955 (SS). Right: 1959 (AL). One-quart cans, $25-$35.

Esso ATF. Left: 1962 (SS). Right: 1965 (SS). One-quart cans, $25-$35.

Esso ATF, 1966 (SS) one-quart can. $25-$35

Esso Marine Oil, 1956 (SS) one-quart can. $30-$60

Esso Marine Engine Oil, Penola Inc. 5/34 (SS) five-gallon can. No estimate.

Left: Esso Outboard, 1955 (SS) one-pint can, $20-$30. Right: Esso Marine Gear Oil, 1955 (SS) one-quart can, $50-$75.

Esso Outboard. Left: 1958 (SS). Right: 1960 (SS). One-quart can, $25-$35.

Esso Motor Cleaner, 1956 (SS) one-quart can. $25-$35

Esso Aquaglide. Left: 1963 (SS). Right: 1965 (SS). One-quart cans, $25-$35.

Esso and Enco Aviation Oil E, 1961 (SS) one-quart cans. $25-$35

Left: Esso Anti-Freeze, 1956 (SS) one-quart can, $25-$35. Right: Esso Tri-Rad, 1950 (SS) one-quart can, $30-$60.

Esso and Enco Aviation Oil E, 1965 (SS) one-quart cans. $25-$35

Esso Aviation Instrument Oil, 1940s (SS), 4 oz. $150-$350

Enco Uniflo, 1969 (SS) one-quart can, $30-$45. Front and back shown.

Enco Uniflo. Left: 1962 (SS). Right: 1964 (SS). One-quart cans, $35-$50.

Enco Extra, 1961 (SS), two variations of the one-quart can. $30-$45

Enco Uniflo. Left: 1965 (SS). Right: 1971 (CP). One-quart cans, $30-$45.

Enco Motor Oil, 1961 (SS), two variations of one-quart can. $30-$45

Enco Motor Oil. Top: 1970 (SS). Left: 1963 (CP). Right: 1965 (SS). One-quart cans, $25-$35.

Encolube. Left: 1964 (CP). Right: 1962 (SS). One-quart cans, $25-$35.

Encolube. Top: 1963 (SS). Left: 1965 (CP). Right: 1966 (SS). One-quart cans, $25-$35.

Encolube HDX. Top: 1961 (SS). Left: 1965 (SS). Right: 1966 (SS). One-quart cans, $25-$35.

Top: Encolube D-3, 1969 (SS) one-quart can. Left: Enco ATF, 1965 (SS). Right: Enco ATF, 1967 (SS). One-quart cans, $25-$35.

Left: Enco Outboard, 1961 (SS) one-quart can, $35-$45. Right: Enco Aquaglide, 1963 (SS) one-quart can, $25-$35.

Carter Lube, 1930s (SS) one-quart can. $150-$350

Oval E Motor Oil, 1945 (SS) one-quart can. $150-$300

Oval E Extra Motor Oil, 1949 (SS), two variations of the one-quart can. $75-$100

Carter Extra, 1950 (SS) one-quart can. $100-$200

Carter Uniflo, 1960 (SS) one-quart can.
$30-$60

Carter ATF, 1960 (SS), one-quart can.
$30-$60

Carter Extra, 1960 (SS), two variations of the one-quart can.
$30-$60

Humble Uniflo. Left: 1955 (SS). Right: 1961 (SS). One-quart cans, $35-$50.

Carter Motor Oil, 1960 (SS), two variations of the one-quart can.
$30-$60

Humble Anti-Freeze, 1955 (SS) one-quart can. $35-$50

Humble 997. Left: 1940 (SS). Right: 1955 (SS). One-quart cans, $35-$100.

Humble Extra, 1971 (SS) one-quart can. $25-$35

Humble Motor Oil, 1971 (SS) one-quart can. $25-$35

Humble Aviation, 1955 (SS) one-quart can. $50-$75

Humble Outboard, 1955 (SS) pint cans. $30-$50

Challenge Motor Oil, 1940s (SS) one-quart can. Pate Oil Co., Milwaukee, WI. $200-$300

Oklahoma Outboard, 1955 (SS) one-quart can. $75-$125

American Petro Fina Inc.

Left: Fina Shield, 1961 (SS) one-quart can, $25-$35. Right: Fina HD Premium, 1961 (SS) one-quart can, $25-$35.

Fiske Bros. Refining Co.

Left: Lubriplate H.D.S. and Lubriplate Non-Detergent. Fiske Bros. R2F Co., 1963 (SS) one-quart cans. $35-$50

Ford Motor Co., Dearborn, MI

Ford Premium, Ford Motor Co., Dearborn, MI. 1975 (CP) one-quart can. $15-$25

Ford Tractor Division, Dearborn, MI

Top: Ford Tractor 15W40 (CP) one-quart can. $10-$25. Left: Ford Tractor 300, (S). Right: Ford Tractor 300, (CP). One-quart-cans, $25-$35.

Ford Tractor Hydraulic Oil. Top: 1965 (SS). Left: 1951 (SS). Right: 1957 (SS). Two-gallon cans, $25-$35 each. Ford Tractor Division, Dearborn, MI

Buick Division of General Motors, Detroit, MI

Left: Buick Anti-Freeze, 1940 (SS) one-quart can, $50-$100. Right: Buick Special Oil, 1951 (SS) one-quart can, $40-$60. Buick Division of G.M.

Cadillac Division of General Motors, Detroit, MI

Left: Cadillac Transmission Oil, 1920 (SS) one-gallon can, $125-$225. Right: Cadillac Hydramatic Fluid, 1956 (SS) one-quart can, $200-$450.

Tidewater Oil Co., New York, NY, Los Angeles, CA

Associated Motor Oil, 1921 (SS) one-gallon can. $325-$675

Veedol Motor Oil, Left to right: 1947 (SS) one-quart can, 1950 (SS) one-quart can, 1952 (SS) one-quart can. $30-$50

Veedol Motor Oil, Left to right: 1954 (SS) one-quart can,1957 (AL) one-quart can, 1959 (AL) one-quart can. $25-$45

Veedol Motor Oil. Top: 1936 (SS) one-quart can. Left: 1938 (SS) one-quart can. Right: 1940 (SS) one-quart can. $50-$75

Tydol. Top: 1940 (SS) one-quart can, Left: 1947 (SS) one-quart can. Right: 1952 (SS) one-quart can. $30-$50

Tydol and Veedol, wartime containers. 1942-1945 one-quart steel cans with plastic type labels wrapped around. $50-$125

Veedol Special. Left: 1961 (SS) one-quart can. Right: 1964 (CP) one-quart can. $25-$35 each.

Veedol 10-30. Left: 1955 (SS) one-quart can. Right: 1959 (SS) one-quart can. $25-$35 each.

Veedol HD+. Top: 1969 (CP). Left: 1961 (SS). Right: 1964 (CP). $25-$35 each.

Veedol 10W30. Left: 1961 (SS) one-quart can. Right: 1964 (CP) one-quart can. $25-$35 each.

Veedol Economy. Left: 1959 (SS). Right: 1964 (CP). $25-$35 each.

Tidex, 1950 (SS) two-gallon can. $25-$35

Veedol ATF A. Left: 1954 (SS) one-quart can. Right: 1959 (AL) one-quart can. $30-$40

Veedol Outboard Sealed Quart, 1961 (SS). $75-$175

Flyng A Anti-freeze. Left: 1948 (SS) one-quart can. Right: 1956 (SS) one-quart can. $30-$40 each.

Veedol Outboard, left to right: 1956 (SS) one-quart can; 1960 (SS) one-quart can; 1962 (SS) one-quart can. $25-$35 each.

Flying A Anti-freeze Coolant, 1965 (SS) one-quart can. $25-$35

Gilmore Oil Co., Los Angeles, California

Lion Head, Gilmore Oil Co., Los Angeles, California. 1940 (SS) one-quart can. $300-$600

Gulfpride, 1934 (SS) one-quart can. $300-$500

Gulf Oil Co., Gulf Building, Pittsburgh, PA

Gulfpride, 1920s (SS) square gallon. $250-$450

Gulfpride, 1930s-1940s (SS) five-quart can. $100-$275

Gulfpride, 1940s (SS) one-quart can. $25-$75

Gulfpride Motor. Top: 4/51. Left: 1949. Right: 4/51 variation. Both (SS) one-quart cans, $25-$50 each.

Gulfpride Alchlor Processed, the World's Finest Motor Oil. 1948 (SS) five-quart can. $150-$200

Gulfpride Motor. Top left: 9/51; top right: 10/59; bottom left: 9/51; bottom right: 2/53. (SS) one-quart cans, $25-$50.

Gulfpride Motor. Top 11/59 (SS). Left: 11/59 (AL). Right: 12/62 (CP), reverse disk on back. $15-$40 each.

Gulfpride H.D. Top: 3/53 (SS). Bottom Left: 1/51 (SS). Bottom Right: 3/55 (SS). One-quart cans, $25-$45.

Gulfpride H.D., both 3/51 (SS) one-quart cans. Note logo variation. $30-$60 each.

Gulfpride Single-G, 10/67 (CP). Left: Gulfpride, 3/68 (CP). Right: Gulfpride, 12/68 (CP). One-quart cans, $15-$30.

Gulfpride H.D. Select. Left: 1/57 (SS). Right: 10/57 (SS). One-quart cans, $25-$40 each.

Gulfpride Select. Top: 11/59 (SS). Left: 11/59 (AL). Right: 12/62 (CP) with reverse disk on back. One-quart cans, $30-$60.

Gulflube, 1930s-1940s five-quart can. (SS) $100-$275

Gulflube, 1930s (SS) square gallon can. $200-$400

Gulflube. Top: 1946 (SS). Right: 1935 (SS). Left: 1940 (SS). One-quart cans, $75-$300 each.

Gulflube, 8/51 (SS) one-quart can. Rare, $50-$75.

Gulflube. Top: 5/60 (SS). Left: 5/60 (AL). Right: 8/62 (CP. One-quart cans, $20-$40.

Gulflube. Left: 3/53 (SS). Right: 1/57 (SS). One-quart cans. $30-$50

Left: Gulflube H-D, 1948 (SS) one-quart can, $75-$150. Right: 2/56 (SS) one-quart can, $50-$75.

Gulflube H-D. Left: 5/60 (SS). Right: 8/62 (CP). One-quart cans, $20-$30.

Gulflube, five-gallon can, 10/63, $50-$75. Rare in one-quart size.

Gulflube. Top: 4/64 (CP). Left: XHD 4/72 (CP). Right: H-D 6/71 (CP). One-quart cans, $15-$25.

Gulf Supreme, 1920s (SS) square gallon can. $250-$450

Gulf Supreme. Left: 1940 (SS). Right: 1946 (SS). One-quart cans, $75-$300 each.

Gulf Supreme. Left: 7/52 (SS). Right: 7/55 (SS). One-quart cans, $40-$60 each.

Gulf Supreme. Left: 5/60 (SS). Right: 5/60 (AL). One-quart cans, $20-$30 each.

Saphire. Left: 8/61 (CP). Right: 8/62 (CP). One-quart cans, $25-$35. Also in two-gallon cans.

Gulf Saphire. Left: 11/63 (CP) one-quart can. Right: 4/69 (CP) one-quart can. $25-$35 each.

Top: Gulfpride Formula G Low Ash, 12/70 (CP). Left: Formula G Motor Oil, 3/68 (CP). Right: Formula G Racing Motor Oil, 3/68 (CP) one-quart cans. $25-$40 each.

Left: Gulfpride Multi-G All Weather, 3/70 (CP) one-quart can. Right: Low Ash, 6/71 (CP) one-quart can. $15-$30 each.

Left: Gulfpride Multi G, 2/72 (CP) one-quart can, $15-$30. Right: Gulfpride Single G, 2/72 (CP) one-quart can. $15-$30 each.

Gulf Super Duty. Left: 12/63 (CP). Right: 9/71 (CP). One-quart cans, $30-$50 each.

Gulf Traffic, 1946 (SS) two-gallon can. $35-$50

Gulf ATF Type A. Top: 5/63 (CP). Left: 9/58 (SS). Right: 5/59 (SS). One-quart cans, $25-$40 each.

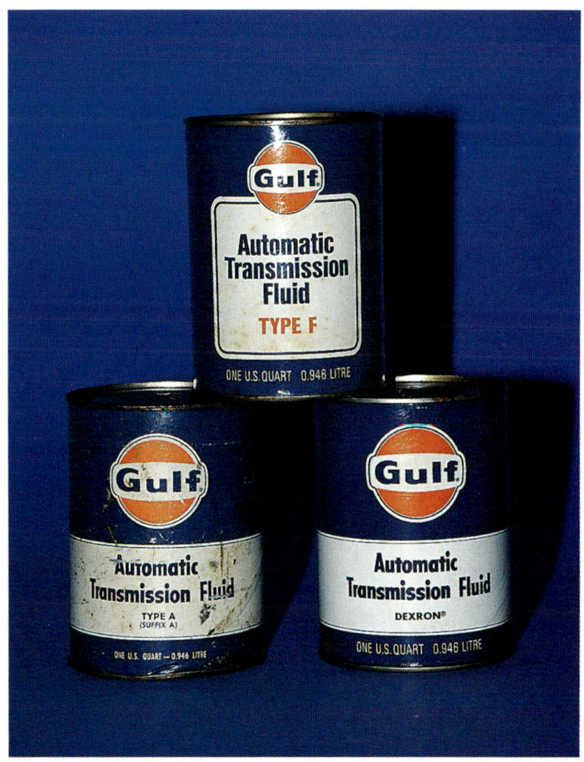

Top: Gulf ATF Type F, 4/69 (CP), $25-$35. Left: Gulf ATF A, 1964 (CP), $25-$35. Right: Gulf ATF Dexron, 4/69 (CP), $25-$35. All one-quart cans.

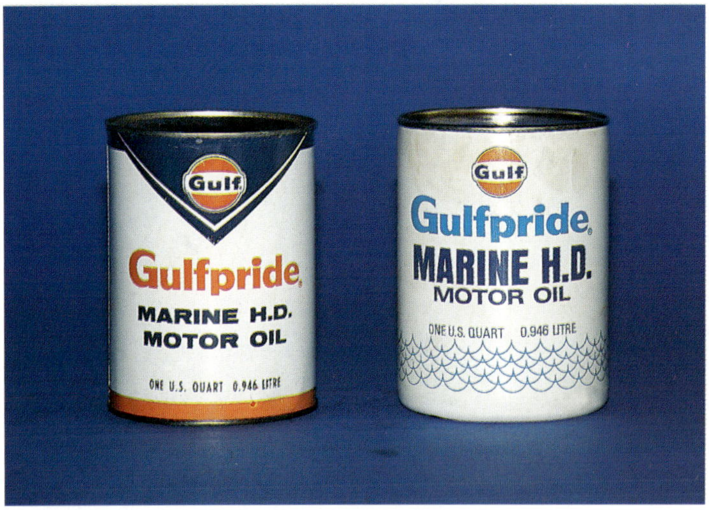

Gulfpride Marine H.D. Left: 4/69 (SS). Right: 12/73 (PC). One-quart cans. $35-45

Left: Gulfpride Aviation, 7/61 (SS), $75-$100. Right: Gulf Aircraft Engine Oil, 8/62 (CP). $50-$75 each.

Gulfpride Aviation. Top: "New", 2/56 (SS). Left: 8/51 (SS). Right: 8/57 (SS). One-quart cans, $75-$100 each.

Gulf Aircraft. Left: 1/57 (SS), Right: 11/59 (SS). One-quart cans, $50-$75 each.

Top: Gulfpride Aviation A-D, 12/73 (P), $35-$35. Left: Gulf Aircraft, 11/63 (PC), $25-$35. Right: Gulfpride Aviation A-D, 4/69 (CP), $25-$35. All one-quart cans.

Gulf Aircraft Oil, 4/74 (PC). One-quart can. $20-$30

Gulf Antifreeze. Left: 1942/1948 (SS). Right: 5/57 (SS). One-quart cans, $50-$75.

Gulf ColdFlo. Left: 1/53 (SS). Right: 2/59 (SS). One-quart cans, $75-$100.

Gulfpride Outboard, left to right: 1939 (SS) one-quart can; 9/55 (SS) one-quart can; 11/59 (SS) one-quart can. $40-$175 each.

Gulf ColdFlo. Left: 5/61 (SS). Right: 1/62 (SS). One-quart cans, $75-$100.

Gulf Outboard Gear Lubricant, 6/51 (SS) one-quart can. $50-$75

Left: Gulf Household Oil, 4/62 (SS) 4 oz. Right: Gulf Lighter Fluid, 10/61 (SS) 4 oz. $35-$50

R.M. Hollingshead, Camden, NJ

Whiz Winter, R.M. Hollingshead, Camden, New Jersey. 1941 (SS) one-quart can. $200-$400

HPX Motor Oil, No company or location. 1950 (SS) one-quart can. $25-$35

Husky Hi Power Inc., Cody, WY

Husky Hi-Lube, 1960 (SS) one-quart can. $150-$350

Husky Heavy Duty, 1957 (SS) one-quart can. $150-$350

Husky, Husky Oil Co., Cody, Wyoming. All 1970 (CP) one-quart cans. $30-$50

Frontier Strato Lube S-3, 1963 (SS). $75-$175

Frontier Oil Co., Denver, CO

Three one-quart Frontier cans dating 1955, all (SS). $100-$300 each.

Frontier/Beeline Strato Lube and Ultra Lube. 1970 (CP) one-quart cans, $25-$40. These cans had both logos.

Three one-quart Frontier cans dating 1963, all (SS). $75-$175

Reverse side of Frontier Strato Lube and Ultra Lube cans.

Frontier/Beeline Antifreeze, 1970 (SS) one-gallon can. $25-$45

Left: Frontier Eterna Grease, 1960 (SS) 35-lb. can, $50-$75. Right: Frontier Lubricant, 1960 (SS) 10-lb. can, $25-$50.

Imperial Refineries, Clayton, MO

Imperial Hep and Imperial Premium, from Imperial Refineries, Clayton, Missouri. Both are 1963 (SS) one-quart cans. $40-$60 each.

International Harvester, Chicago, IL

Right to left: Inter Season 1973 (CP) one-quart can; No. 1 1973 (CP) one-quart can; Touch Control 1961 (SS) one-quart can. $25-$35 each.

Deep Rock Oil Co., Kerr McGee Oil Co., Oklahoma City, OK

Deep Rock Air Race, 1940 (SS) one-quart can. $150-$300

Deep Rock Prize, 1946 (SS) one-quart can. $40-$60

Deep Rock cans. All 1958 (SS) one-quart cans. $35-$150 each.

Left to right: Deep Rock HD, 1961 (SS) one-quart can. Deep Rock Valu Lube, 1961 (AL) 1 quart can. Deep Rock Prize, 1960 (AL) one-quart can. $20-$30 each.

Deep Rock cans. 1969 (CP) one-quart cans. $25-$35

Deep Rock, 1940 (SS) two-gallon can. $35-$55

Lion Oil Co., Eldorado, AR

Lion Oil Co., Eldorado, AR. Left: 1963 (SS). Right: 1966 (CP). One-quart cans, $40-$60.

Lion Oil Co., Eldorado, AR. All 1969 (CP) one-quart cans. $20-$30

Marathon Oil Co., Findley, OH

Left: Marathon Extended Life, 1965 (CP) one-quart can, rare, $75-$150. Right: Marathon Superior All Season, 1978 (CP) one-quart can, $10-$20.

Marathon All Season. Left: 1962 (SS) one-gallon can. Right: 1966 (SS) 1-gallon can. $25-$35 each.

Marathon All Season. Top: 1978 (CP). Left: 1971 (CP). Right: 1973 (CP). One-quart cans, $15-$25.

Left: Marathon VEP 10W30, 1960 (SS) one-quart can. Right: Marathon VEP, 1948 (SS), one-quart can. $50-$125

Marathon Endurance. 1957 (SS) two-gallon can. $40-$75

Marathon VEP. Top: 1978 (CP). Left: 1962 (S). Right: 1965 (CP). One-quart cans, $25-$35 each.

Left: Marathon Multi Power S-3, 1974 (CP) one-quart can. Right: Marathon Ultra D, 1980 (CP) one-quart can. $25-$35

Marathon Endurance. Left: 1947 (SS) one-quart can. Right: 1959 (SS) one-quart can. $50-$125 each.

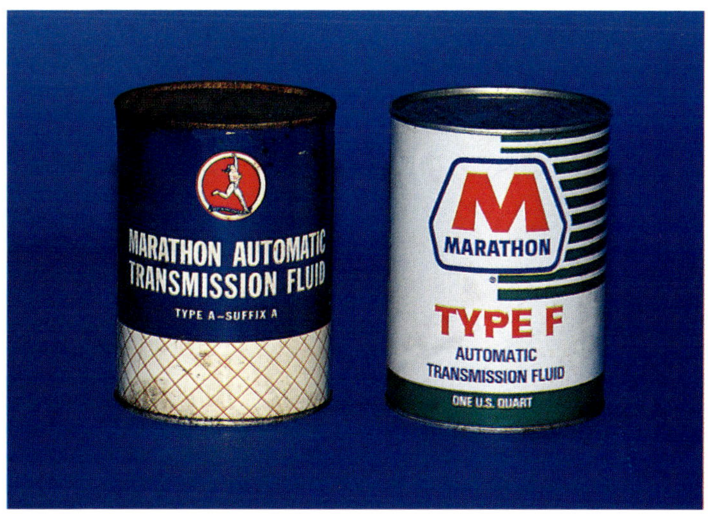

Left: Marathon ATF A, 1960 (SS) one-quart can. Right: Marathon ATF F, 1978 (CP) one-quart can. $15-$35 each.

Marathon Anti-Freeze. Left: 1964 (SS) 1-gallon can. Right: 1973 (PC) one-quart can. $15-$30 each.

Marathon Marine Master, 1965 (SS) one-pint can. $15-$25

Marathon Outboard, 1958 (SS) one-quart can. $35-$45

Marathon. Left: 1962 (SS) 4-oz. can. Right: 1965 (SS) 4-oz. can. $15-$25 each.

Gastown Division of Marathon Oil Co.

Gastown, 1962 (S) two-gallon can. $20-$30

Minute Man Service Inc., Lakewood New York

Minute Man Motor Oil, Minute Man Service Inc., Lakewood New York. 1963 (SS) two-gallon can. $30-$50

Mobil Oil Co., New York, NY

Left: Socony Motor Oil, 1922 (SS) one-quart can. Right: Socony Cup Grease, 1922 (SS) one-pound can. $50-$225

Mobiloil E. 1929 (SS) one-gallon can. $125-$225

Socony Kerosene Oil. 1925 (SS) one-gallon can. $175-$375

Mobiloil Arctic Special, 1937 (SS) one-quart can. $75-$150

Mobiloil Gargoyle, Socony-Vacuum. All 1935 (SS) one-quart cans. $125-$225 each.

Mobiloil Gargoyle, both 1941 (SS) one-quart cans. $50-$75 each. This vintage began the series of Mobiloil cans with the type of oil stamped on top.

Mobiloil Gargoyle, Socony-Vacuum. Both 1935 (SS) one-quart cans. $125-$225

Mobiloil cardboard cans. 1942-1945 one-quart. $50-$75 each. These were used during World War II to conserve steel for the war effort.

Mobiloil Super, 1964 (CP) one-quart can. $25-$40

Mobiloil Special. Left: 1963 (SS), Right: 1965 (CP). One-quart cans. $25-$35 each.

Mobiloil Special. Left: 1956 (SS). Right: 1959 (SS). One-quart cans, $35-$45.

Mobiloil, 1946 (SS) one-quart cans. Three top variations. $30-$40 each.

Mobiloil Special, 1961 (AL) one-quart cans. Two variations. $25-$35 each.

Mobiloil, 1948 (SS) one-quart cans. Two top variations. $30-$40 each.

Mobiloil, 1951 (SS) one-quart cans. Two top variations. $25-$35 each.

Mobiloil, 1953 (SS) one-quart cans. Two top variations. $25-$35 each.

Mobiloil, 1955 (SS) one-quart cans. Two top variations. $25-$35

Mobiloil, 1961 (AL) one-quart cans. Two top variations. $25-$35

Mobiloil, 1962 (SS) five-quart can, $40-$60. Also sold in one-quart size.

Mobiloil, 1962 (AL) one-quart cans. Five top variations. $25-$35 each.

Mobiloil, 1963 (SS) one-quart cans. Two variations. $25-$35 each.

Mobiloil, 1965 (CP) one-quart can. $20-$30

Magnolia Lubrite, 1940 (SS) one-quart can. $50-$75

Left: Standard Lubrite, 1940 (SS) one-quart can. Right: Lubrite, 1940 (SS) one-quart can. $30-$50 each.

Standard Lubrite, 1946 (SS) one-quart can. $30-$50

Lubrite. Left: 1951 (SS). Right: 1953 (SS). One-quart cans, $25-$35 each.

Lubrite, 1946 (SS) one-quart cans. Two top variations with tops shown. $30-$50

Lubrite, 1955 (SS) and 1956 (AL) one-quart cans. $30-$40 each.

Lubrite. Left: 1958 (SS). Right: 1960 (AL). One-quart cans, $25-$35 each.

Mobiloil Outboard, 1956 (SS) one-quart can. $125-$225

Mobil Delvac. Top: 1965 (AL); information is stamped on the top. Left: 1956 (SS). Right: 1959 (AL). One-quart cans. $50-$125

Mobiloil Outboard Special, 1958 (SS) one-quart can. $125-$225

Mobiloil Outboard, 1962 (SS) one-quart can. Variation in diamonds. $30-$50

Mobiloil Marine, 1956 (SS) one-gallon can. $50-$75

Mobiloil Marine, 1958 (SS) one-quart can. $150-$350

Mobilgrease UW. 1948 (SS) one-quart can. $125-$175

Mobil Marine Grease, 1956 (SS) one-pound can. $50-$75

Mobiloil, 1934 (SS) three-gallon can. $90-$125

Mobiloil Marine, 1955 (SS) one-quart can. $40-$60

Mobil three-gallon can, 1940 (SS). $75-$150

Mobil Metro, 1964 (SS) two-gallon can.
Mobil Lubrite, 1960 (SS) two-gallon can. $35-$60

Mobilfluid. Left: 1953 (SS). Right: 1955 (SS). One-quart cans, $30-$50.

Mobilfluid, 1964 (CP) one-quart can. Rare, $30-$50.

Mobil Freezone. Left: 1938 (SS). Right: 1941 (SS). One-quart cans, $30-$50 each.

Mobilfluid. Top: 1961 (AL). Left: 1960 (AL). Right: 1962 (SS) one-quart cans. $30-$50

Mobil Freezone. Left: 1953 (SS). Right: 1955 (SS). $30-$45 each.

Mobil Freezone. Left: 1957 (SS). Right: 1959 (SS). One-quart cans, $30-$40 each.

Mobil Permazone. Top: 1963 (SS). Left: 1957 (SS). Right: 1959 (SS). One-quart cans, $30-$40 each.

Left: Mobiloil Aero Red Band, 1946 (SS). Right: Mobiloil Green Band, 1946 (SS). One-quart cans, $125-$225 each.

Mobiloil Aero Gray Band, 1946 (SS) one-quart can. $125-$225

Mobiloil Aero Red Band, 1957 (SS) one-quart can. $30-$40

Mobiloil Aero, 1951-1953, all (SS). Three variations. One-quart cans, $30-$40 each.

Mobiloil Aero Red Band, 1961 (AL) one-quart can. 25-$35

Mobil Outboard, 1947-1960. All (SS), all one-quart cans. $20-$75 each.

Mobil Lube Outboard Gear Oil, left to right: 1947 (SS) one-quart can. 1950 (SS) one-quart can. 1955 (SS) one-quart can. $35-$125 each.

Socony Household Oil. Left: 1925 (SS) 4 oz. Right: 1931 (SS) 4 oz. $100-$250

Mountain State Motor Oil

Mountain State Motor Oil. No company name or location. 1930s (SS) two-gallon can. $50-$75

Murphy Oil Co., Eldorado, AR

Spur, Murphy Oil Co., Eldorado, AR. Left: 1982 (CP) one-quart can. Right: 1982 (CP) one-quart can. $25-$35

National Refining Co. Oklahoma City, OK

EN-AR-CO, National Refining Co. Oklahoma City, Oklahoma. All 1960 (SS) one-quart can. $30-$60

Nourse Oil Co., Kansas City, MI; Omaha, NE

Neatslene Harness Oil, Nourse Oil Co., Kansas City, Missouri; Omaha, Nebraska. 1920s (SS) one-gallon can. $100-$225

Northland Products Co., Waterloo, IA

Northland Motor Oil, Northland Products Co., Waterloo, Iowa. Left: 1961 (SS). Right: 1964 (CP). One-quart cans, $30-$50.

Phillips Petroleum Co., Bartlesville, OK

Phillips Trop-Artic. Left: 1936 (SS). Center: 1945 (SS). Right: 1950 (SS). One-quart cans, $50-$175.

Phillips 66 Trop-Artic. Left: 1955 (SS). Right: 1959 (AL). One-quart cans, $25-$35 each.

Phillips 66 Trop-Artic. Left: 1960 (SS). Right: 1962 (AL). One-quart cans, $25-$35 each.

Phillips 66 Motor Oil. Left: 1931 (SS). Right: 1934 (SS). One-quart cans, $50-$75 each.

Phillips 66 Sixty-Six. Left: 1960 (SS). Right: 1965 (CP). One-quart cans, $20-$30 each.

Phillips 66 Motor Oil. Left to right: 1936, 1940, 1945, 1955. All (SS) one-quart cans, $25-$50 each.

Phillips Sixty Six Special, 1965 (CP) one-quart can, $25-$35.
Phillips 66 MM, 1965 (CP) one-quart can, $25-$35.

Phillips 66 Premium. Left: 1950 (SS). Right: 1959 (SS). One-quart cans, $25-$45 each.

Phillips 66 Unique. Left to right: 1955 (SS), 1959 (AL),1960 (SS), 1965 (CP). One-quart cans, $25-$75.

Phillips 66 Super HD, 1963 (AL) one-quart can. $30-$40

Phillips 66 Racing, 1970 (CP) one-quart can. $25-$35

Philube ATF A. Left: 1951 (SS). Center: 1959 (AL). Right: 1965 (CP). One-quart cans, $25-$400 each.

Phillips 66 HDS, Left to right: 1954 (SS), 1958 (SS), 1960 (SS), 1965 (CP). One-quart cans, $25-$75.

Phillips 66 Aviation, left to right: 1955 (SS), 1960 (SS), 1963 (AC), 1965 (CP). One-quart cans, $25-$75 each.

Phillips 66 Marine HD, 1956 (SS) one-quart can. $200-$400

Phillips 66 Outboard. Left: 1959 (SS) one-quart can. Right: 1954 (SS) one-quart can. $25-$45 each.

Left: Phillips 66 Marine HD, 1961 (AL) one-quart can, $200-$400.
Right: Phillips 66 Outboard, 1965 (PB), $10-$20.

Phillips 66 Outboard, left to right: 1955 (SS) one-quart can. 1960 (SS) one-quart can. 1962 (SS) one-quart can. $25-$45 each.

Phillips 66 Anti-Freeze. Left: 1948 (SS). Right: 1957 (SS). One-gallon can, $35-$100 each.

Left: Phillips 66 Cooling System Fluid, 1963 (SS) one-quart can, $25-$35. Right: Phillips 66 Anti-Freeze, 1964 (SS) one-quart can, $25-$35.

Phillips 66 Anti-Freeze, 1957 (SS) one-quart can. $35-$45

Phillips 66 Anti-Freeze. Left: 1960 (SS). Right: 1963 (SS). One-quart cans, $25-$35 each.

Pittsburgh Penn Oil Co., Pittsburgh, PA

Zero Pruf, Pittsburgh Penn Oil Co., Pittsburgh, Pennsylvania. 1961 (SS) one-quart can. $50-$75

Penn Refining Co., Butler, PA; Cleveland, OH; Edgewater, NJ; Los Angeles, CA

Penn Drake, Penn Refining Co., Butler, Pennsylvania; Cleveland, Ohio; Edgewater, New Jersey; Los Angeles, California. 1963 (SS) one-quart can. $15-$25

Colonial Drake HD, Penn Refining Co., Butler, Pennsylvania. 1960 (SS) one-quart can. $25-$35

Penn-Super Products Co., USA

Penn Super, Penn-Super Products Co., USA. 1938 (SS) two-gallon can. $30-$50

Publicker Industries, Philadelphia, PA

Thermo Anti-Freeze, Publicker Industries, Philadelphia, Pennsylvania. 1945 (SS) one-quart can. $75-$175

Penzoil Oil Co., Oil City, PA

Penzoil, Penzoil Oil Co., Oil City, Pennsylvania. 1946 (SS) one-quart can. $50-$75

Pure Oil Co., Chicago, IL; Palatine, IL

Purol, 1940 (SS) one-quart can. $30-$60

Pure Tiolene. Left: 1938 (SS), Pennsylvania oil logo on side of can. Right: 1940 (SS), Aviation quality with airplane on side of can. $40-$70 each.

Pure Tiolene/Lube. Left to right: 1945 (SS), 1949 (SS), 1951 (SS), 1951 top variation (SS). One-quart cans, $35-$65 each.

Purelube. 1954 is not pictured (SS). Left to right: 1955 (SS), 1956 (SS), 1960 (SS), 1962 (CP). One-quart cans, $35-$65.

High HP Purelube, front and back are pictured. Left: 1954 (SS). Right: 1955 (SS) One-quart cans, $25-$45 each.

Puritan, Left to right: 1953 (SS), 1955 (SS), 1960 (SS), 1962 (CP). One-quart cans, $25-$45.

High HP Purelube. Left: 1960 (SS). Right: 1962 (CP). One-quart cans, $20-$45.

Pure Guardian, 1962 (CP) one-quart can. $25-$45

Pure Guardian, 1930s-1940s (SS) two-gallon can. $30-$50

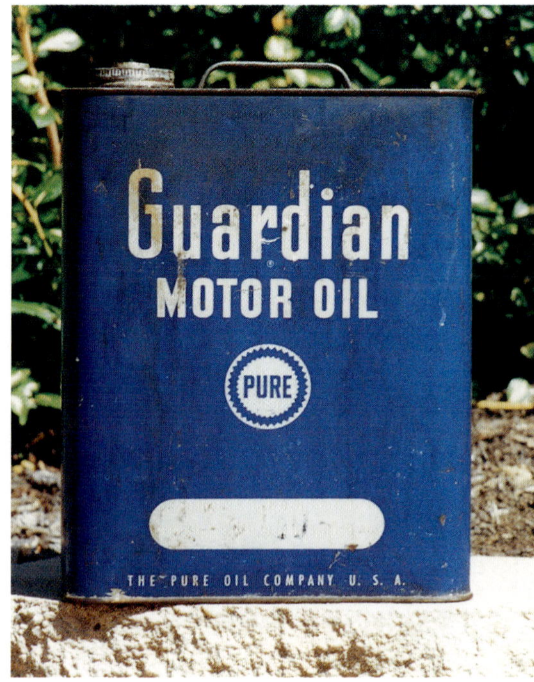

Pure Guardian (silver letters), 1957 (SS) two-gallon can. $25-$35

Pure Guardian. 1954 (SS) two-gallon can. $30-$60

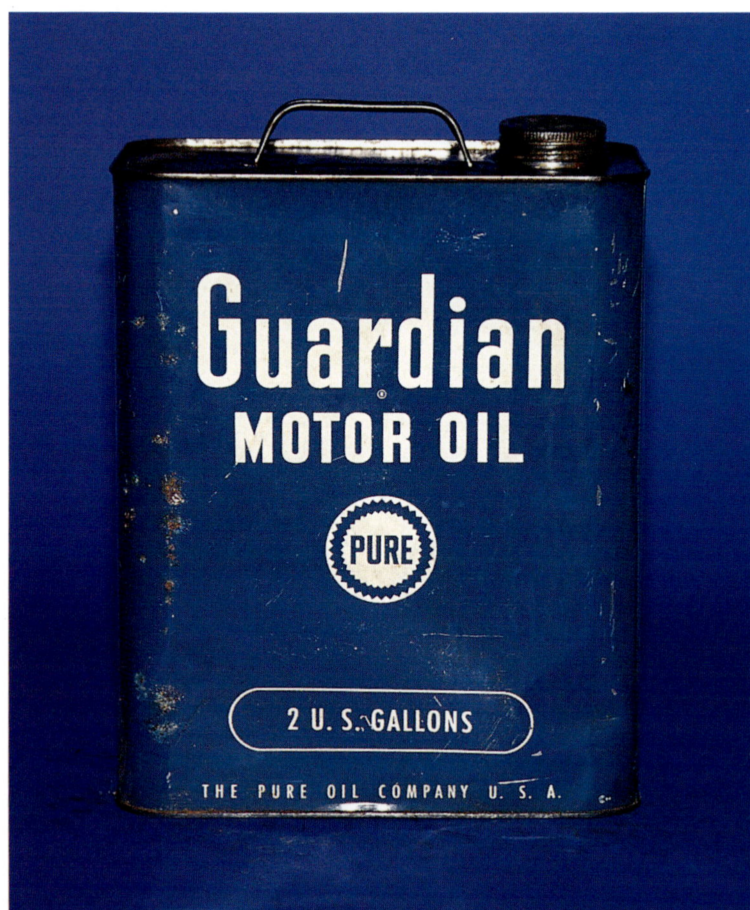

Pure Guardian (white letters), 1960 (SS) two-gallon can. $25-$35

Pure Citation. Left: 1962 (CP). Right: 1964 (CP). One-quart cans, $25-$40.

Pure Motor Oil, 1965 (CP) one-quart can. $25-$35

Left: High HP Purelube, 1965 (CP) one-quart can. Right: Pure Super Duty, 1965 (CP) one-quart can. $25-$35

Purol Heavy Duty. 1951 (SS) and 1953 (SS) one-quart cans. $30-$60 each.

Purol. Left to right: 1957 (SS), 1959 (SS), 1962 (CP) one-quart cans. $30-$60 each.

Purol, All dated 62 with top and back variations. One-quart cans, $25-$35 each (CP)

Purol Super Heavy Duty. Left: 1966 (CP), Right: 1963 (CP). One-quart cans, $35-$50.

Hi-Speed Solvenoil, Hi-Speed Oil Co., Hickock Oil Co., Toledo Ohio. 1945 (SS) one-quart can, $125-$225. Pure Oil Company bought out Hi-Speed in 1951.

Purelube ATF A. Top: 1966 (CP). Left: 1955 (SS). Right: 1962 (CP). One-quart cans. $25-$40

Pureflight Aviation. Top: 1966 (CP). Left: 1955 (SS). Right: 1962 (CP). One-quart cans, $40-$175.

Left: Pure One-Fill Anti-Freeze, 1953 (SS) one-gallon can, $50-$100. Right: Pure Safety Anti-Freeze, 1949 (SS) one-gallon can, $100-$300. Also sold in quarts.

Pure One-Fill. Top: 1962 (SS). Left: 1960 (SS). Right: 1966 (SS). One-quart cans, $30-$50 each.

Pure-sure Anti-Freeze. Top: 1955 (SS). Left: 1953 (SS). Right: 1954 (SS). One-quart cans, $30-$50 each.

Purelube Outboard. Left: 1955 (SS) one-quart can. Right: 1961 (SS) one-quart can. $35-$55 each.

Purelube Outboard, 1962 (CP) one-quart can. $25-$50

Pure Oil. Left to right: 1938 (SS) 1-1/2 oz.; 1950 (SS) 4 oz.; 53-1 (SS) 4 oz.; 1967 (SS) 4 oz.; 62-4 (SS) 4 oz. $40-$200 each.

Pure-Tect Rust Proof, metal can with paper label, one-quart can. $25-$35

Red Head Oil Co.

Red Head Motor Oil/ATF, Red Head Oil Co. 1961 (SS) one-quart can. $25-$35

Red Raven Oil Co., Canton, OH

Pennoleum, Red Raven Oil Co., Canton, Ohio. 1920s (SS) one-gallon can. $75-$125

Roosevelt Oil Co., Mt. Pleasant, MI

ROC Motor Oil, Roosevelt Oil Co., Mt. Pleasant, Michigan. 1950 (SS) one-quart can. $40-$60

Shamrock Oil & Gas Co., Amarillo, TX

Shamrock LP Gas, 1958 (SS) one-quart can. $75-$125

Shamrock ECOlube. Left: 1965 (SS) one-quart can, $30-$50. Right: 1970 (CP) one-quart can, $25-$35.

Shamrock Equa Flow. Left: 1964 (SS). Right: 1970 (CP). One-quart cans, $25-$35.

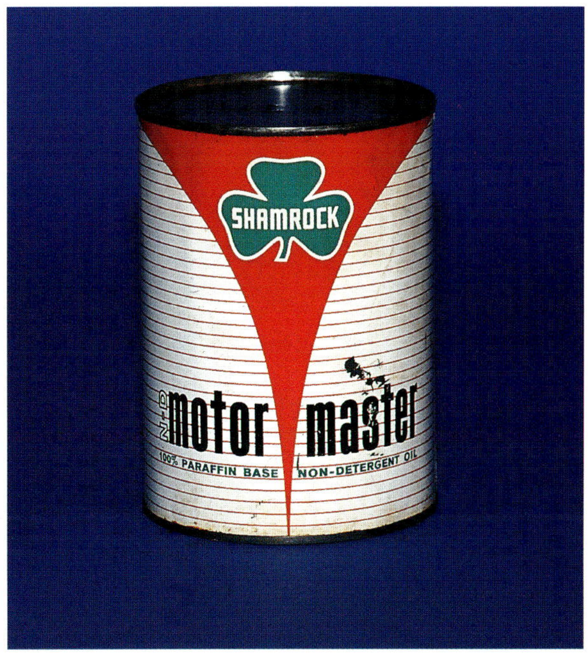

Shamrock Motor Master, 1965 (SS) one-quart can. $30-$50

Shamrock Triple Action. Top: 1970 (CP). Left: 1965 (SS). Right: 1965 S-3 Oil (SS). One-quart cans, $25-$35.

Shamrock Household Oil. Left to right: 1955 (SS) 4 oz.; 1960 (SS) 4 oz.; 1964 (SS) 4 oz. $30-$40

Shell Oil Co., New York, NY; San Francisco, CA

Top: Shell Penn, 1940 (SS) one-quart can, $250-$500. Left: Shell Zone Anti-Freeze, 1945 (SS) one-quart can, $50-$75. Right: Golden Shell, 1960 (SS) one-quart can, rare, $35-$50.

Shell X100. 1956 (SS) and 1954 (SS) one-quart cans. $25-$35

Shell X100. Top: 1951 (SS). Left: 1941 (SS). Right: 1946 (SS). One-quart cans, $25-$40.

Shell X100. Left: 1960 (SS). Right: 1957 (SS). One-quart cans, $25-$35.

Shell X100. Top: 1971 (CP). Left: 1969 (CP). Right: Shell X100 Multigrade, 1969 (CP). One-quart cans, $15-$25 each.

Shell X100 Premium. 1955 and 1956 (SS) one-quart cans. $25-$50

Shell X100 Premium. 1957 (RS) and 1960 (SS) one-quart cans. $25-$35

Golden Shell. Top: 1951 (SS). Left: 1938 (SS). Right: 1940 (SS). One-quart cans, $35-$250.

Golden Shell. Top: 1956 (SS). Left: 1961 (SS). Right: 1969 (CP). One-quart cans, $25-$40.

Super Shell, 1967 (SS) one-gallon can, $30-$50. Also sold in one-quart cans.

Silver Shell. Left: 1941 (SS) two-gallon can. Right: 1960 (SS) two-gallon can. $25-$175 each.

Shell Rotella. Left: 1953 (SS). Right: 1956 (SS). One-quart cans, $25-$35 each.

Silver Shell. 1960 (SS) and 1963 (SS) two-gallon cans. $25-$35 each.

Shell Rotella.
Top: 1969 (CP).
Left: 1957 (SS).
Right: 1961 (SS). One-quart cans, $30-$40.

Rotella T Oil. Left: 1962 (SS), one-quart can, rare, $30-$50. Right: 1969 (CP) one-quart can, $25-$35.

Shell Donax T-6. Top: 1966 (CP). Left: 1956 (SS). Right: 1961 (SS). One-quart cans, $35-$50.

Shell Outboard. Left: 1956 (SS) one-quart can. Right: 1963 (SS) one-quart can. $25-$45 each.

Shell Outboard resealable round quart. Left: 1953 (SS). Right: 1956 (SS). $50-$75 each.

Shell Outboard sealed quart, 1963 (SS). $50-$125

Shell Engine Conditioning Oil. Top: 1967 (CP). Left: 1956 (SS). Right: 1961 (SS). One-quart cans, $25-$45 each.

Aeroshell. Top: 1969 (CP). Left: 1961 (SS). Right: 1957 (SS). One-quart cans, $25-$40.

Aeroshell Premium. Left: 1961 (SS). Right: 1969 (CP). One-quart cans, $25-$40.

Aeroshell. Left: 1945 (SS). Right: 1955 (SS). One-quart cans, $30-$50.

Shell Anti-Freeze, 1941 (SS) one-gallon can. $50-$75

Shell Anti-Freeze. Top: 1940 (SS). Left: 1955 (SS). Right: 1948 (SS). One-quart cans, $45-$100.

Shell Anti-Freeze. Top: 1958 (SS). Left: 1957 (SS). Right: 1960 (SS). One-quart cans, $30-$60.

Top: Shellzone, 1961 (SS). Left: Shell Anti-Freeze, 1950 (SS). Right: Shellzone, 1953 (SS). One-quart cans, $30-$60.

Shellzone. Top: 1966 (SS). Left: 1962 (SS). Right: 1963 (SS). One-quart cans, $30-$50.

Shellzone. Left: 1969 (SS). Right: 1971 (CP). One-quart cans, $25-$35.

Signal Oil Co., Los Angeles, CA

Signal Oil Co., Los Angeles, California. Left: 10W30, 1965 (CP). Right: RPM Delo, 1965 (CP). One-quart cans, $75-$150 each.

Hancock, Mfg. for Signal Oil & Gas Co., Los Angeles, California and Houston, Texas. 1969 (CP) one-quart can. $50-$75

Sinclair Refining Co., NY

Pennant Motor Oil, Pierce Petroleum Co., St. Louis, Missouri. 1920s (SS) one-quart can. No estimate. This company was bought by Sinclair in the 1920s.

Sinclair Opaline Motor Oil. 1920s (SS) one-gallon can. $175-$375

Sinclair Motor Oil. 1920s (SS) one-gallon can. $175-$375

Top: Sinclair Pennsylvania, 1935 (SS) one-quart can, $45-$65. Bottom: Sinclair Opaline, 1935-1945 (SS) one-quart cans, $45-$65.

Left: Sinclair Opaline, 1934 (SS) one-quart can, $50-$100. Right: Sinclair Pennsylvania, 1934 (SS) one-quart can, $100-$250

Top: Sinclair Opaline, 1953 (SS). Left: Sinclair Pennsylvania, 1953 (SS). Right: Sinclair Pennsylvania HD, 1953 (SS). One-quart cans, $25-$50.

Sinclair Triple-X. Left: 1955 (SS). Right: 1958 (SS). One-quart cans, $25-$45.

Sinclair Triple-X. Left: 1961 (SS). Right: 1964 (CP). One-quart cans, $25-$35 each.

Sinclair Extra Duty. Left: 1955 (SS). Right: 1958 (SS). One-quart cans, $25-$45 each.

Sinclair Extra Duty. Left: 1961 (SS). Right: 1961 (CP). One-quart cans, $25-$35 each.

Sinclair Opaline. Top: 1955 (SS). Left: 1961 (SS). Right: 1964 (CP). One-quart cans, $25-$35 each.

Sinclair Pennsylvania. Left: 1955 (SS). Right: 1961 (SS). One-quart cans, $25-$35 each.

Sinclair DTR. Right: 1969 (CP) one-quart can, $50-$75. Left: Sinclair Dino Track & Road, 1971 (CP) one-quart can, $75-$150.

Sinclair Tenol. Top: 1958 (SS). Left: 1961 (SS). Right: 1964 (CP). These cans came with colored tops to separate the different grades, starting with 1961 vintage. These colors are blue, red, green, yellow. One-quart cans, $25-$35.

Sinclair Tenol oils, both 1969 (CP). Colors: blue, red, black, green, white, yellow. One-quart cans, $25-$35.

Sinclair Emerald. Left: 1947 (SS). Right: 1965 (SS). Two-gallon cans, $35-$75.

Sinclair Emerald, 1956 (SS) two-gallon can. $25-$35

Sinclair Marine HD, 1956 (SS) one-quart can. $75-$125

Sinclair Extra Duty Outboard, 1958 (SS) one-quart can. $50-$75

Sinclair Emerald, 1963 two-gallon can (SS). $25-$35

Left: Sinclair Outboard Special, 1959 (SS) one-quart can, $50-$75. Right: Extra Duty Outboard, 1962 (SS), 1964 (CP) one-quart can, $25-$75.

Sinclair Outboard. Left to right: 1958 (SS), 1959 (SS), 1962 (SS). One-quart cans, $30-$40 each.

Sinclair Outboard Gear Oil, 1956 (SS) one-quart can. $40-$60

Sinclair Aircraft, 1956 (SS) one-quart can. $50-$125

Top: Sinclair AVOIL, 1972 (CP) one-quart can, $25-$50. Left: Sinclair Aircraft, 1962 (SS). Right: Sinclair Aircraft, 1964 (CP). One-quart cans, $25-$75.

Sinclair Aircraft. Left: 1961 (SS). Right: 1963 (SS). One-quart cans, $25-$50.

Top: Sinclair Engine Purge, 1963 (SS) one-quart can, $25-$35. Left: Diesel Fuel Additive A, treats 1450 gals 1965 (SS), one-quart can, $50-$75. Right: Diesel Fuel Additive B, treats 225 gals 1965 (S), one-quart can, $50-$75.

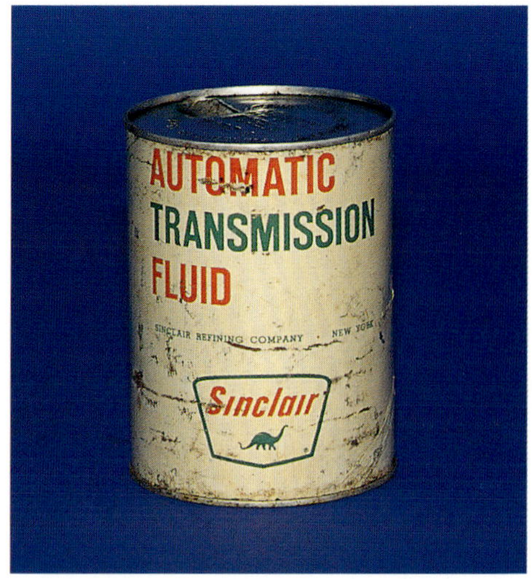

Sinclair ATF Type A, 1963 (CP) one-quart can. $25-$35

Sinclair Anti-Freeze, 1955 (SS) one-gallon can. $50-$75

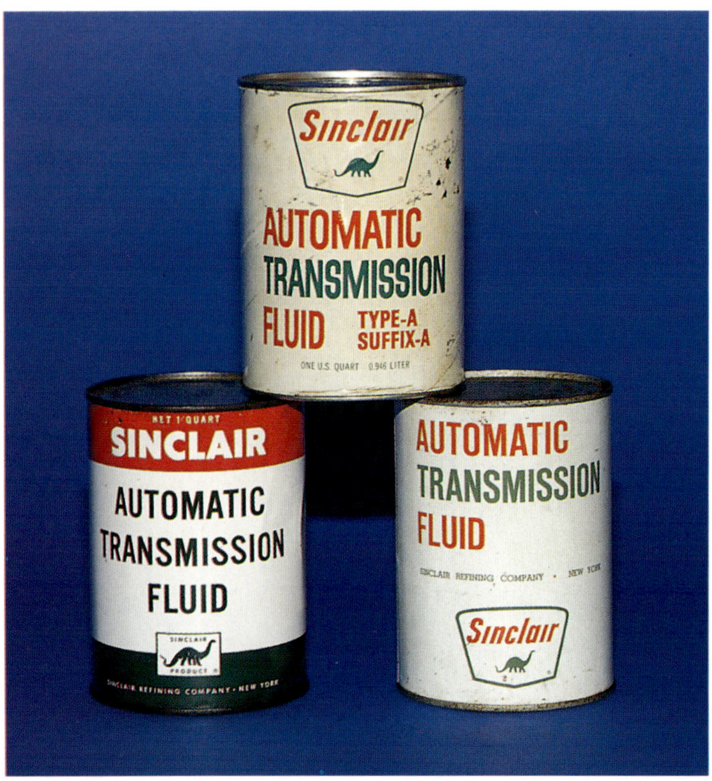

Sinclair ATF Type A. Top: 1969 (CP). Left: 1957 (SS). Right: 1961 (SS). One-quart cans, $25-$35 each.

Sinclair Anti-Freeze. Left: 1957 (SS). Right: 1966 (SS). One-quart cans, $25-$35.

Skelly Oil Co., Tulsa, OK

Skelly UniFlo, 1935 (SS) one-quart can. $25-$35

Skelly Supreme. Left: 1960 (SS). Center: 1963 (RS). Right: 1969 (RS). One-quart cans, $25-$35.

Left: Skelco, 1950 (SS) one-quart can. Right: Skelly Universal, 1946 (SS) one-quart can. $25-$40 each.

Skelly Tagoline/Tagoline HD. Left: 1940 (SS). Center: 1947 (SS). Right: 1951 (SS). $25-35

Skelly Supreme. Left: 1947 (SS). Center: 1955 (SS). Right: 1958 (SS). One-quart cans, $30-$40.

Skelly Tagoline HD. Left: 1958 (SS). Center: 1963 (RS). Right: 1969 (RS). One-quart cans. $25-$35

Skelmark. Left: 1960 (SS). Right: 1963 (RS). One-quart cans, $35-$45.

Skelly Ranger. Left to right: 1952 (SS), 1958 (SS), 1963 (RS), 1969 (RS). One-quart cans, $25-$35 each.

Skelly Motor Oil. 1954 (SS) one-quart can. $25-$35

Skelly S-3. Left to right: 1958 (SS), 1961 (RS), 1963 (RS), 1969 (RS). One-quart cans, $35-$45.

Skelly Motor Oil. Left: 1963 (RS). Right: 1969 (RS). One-quart cans, $25-$35.

Skelly Commercial Oils, 1963-1969 (RS) one-quart cans. These cans are plain white with the tops used for product ID. $25-$35 each.

Left: Skelly Heavy Duty Truck HDT 100, 1973 (CP). Right: Skelly HDT 200/300, 1973 (CP). One-quart cans, $15-$25.

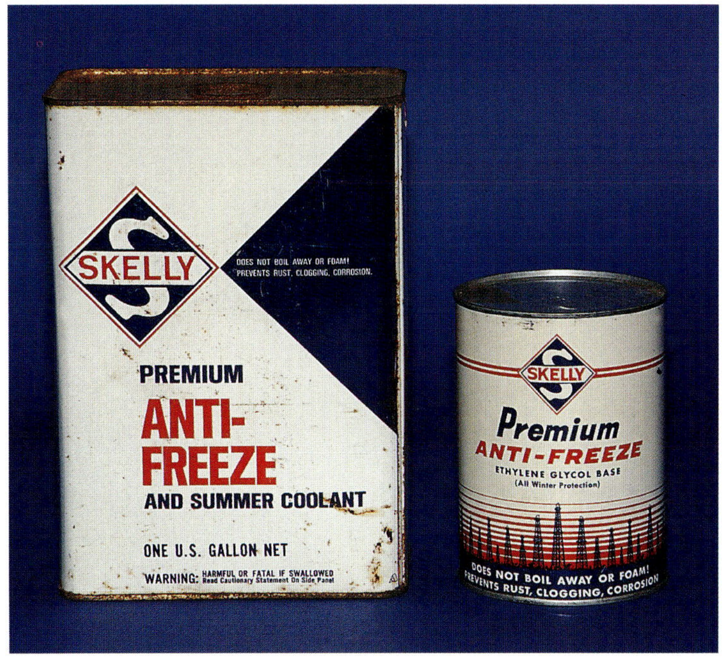

Skelly Premium Anti-Freeze. Left: 1969 (SS) one-gallon can, $25-$35. Right: 1963 (SS) one-quart can, $50-$90.

Skellmatic ATF A. Left to right: 1956 (SS), 1963 (RS), 1969 (RS). One-quart cans, $25-$35.

Skelly Anti-Freeze, 1961 (SS) one-gallon can. $25-$35

Skelly Supreme Outboard cone top can (SS), 1960, one-quart. $35-$45

Standard Oil Co. of Ohio, Cleveland, OH

Polarine Motor Oil, Standard Oil Co. Inc. 1910 (SS) one-gallon can. $375-$1200

Sohio Motor Oil. 1930 and 1938 (SS) one-quart tall cans. $150-$225 each.

Sohio Motor Oil. Left to right: 1938 (SS), 1940 (SS), 1946 (SS). One-quart cans, $35-$65.

Sohio Motor Oil. Left to right: 1955 (SS), 1960 (SS), 1962 (SS). One-quart cans, $20-$40.

Standard Oil Penn. Left: 1934 (SS) five-quart can. Right: 1938 (SS) five-quart can. Also sold in one-quart cans. $75-$150 each.

Sohio HQ. Top: 10/47 (SS). Left: 2/47 (SS). Right: 3/47 (SS). One-quart cans, $75-$300.

Sohio HQD. Top: 1962 (SS). Left: 1964 (SS). Right: 1966 (SS). One-quart cans, $30-$40.

Sixty Two, 1967 (SS) one-quart can. $50-$100

Sohio HQD. Top: 1960 (SS), Left: 1955 (SS), Right: 1951 (SS). One-quart cans, $35-$65.

Premex. Top: 1962 (SS). Left: 1955 (SS). Right: 1957 (SS). One-quart cans, $30-$60 each.

Sohio Premex. Left to right: 1964 (SS), 1965 (SS),1967 (SS), 1969 (SS). One-quart cans. $25-$35

Sohio Duron. Left: 1962 (SS). Right: 1964 (SS). One-quart cans, $25-$50.

Sohio QVO. Top: 1972 (SS). Left: 1968 (SS). Right: 1969 (SS). One-quart cans, $25-$35.

Sohio Golden Duron. Top: 1965 (SS). Left: 1969 (SS). Right: 1973 (SS). One-quart cans, $25-$35.

Sohio Nitrex. Left: 1976 (SS). Right: 1981 (SS). One-quart cans, $20-$35.

Polarine, 1962 (SS) one-quart can. $100-$200 (Rare)

Left: Sohio Marine Engine Oil, 1956 (SS) one-quart can, $125-$250. Right: Sohio 7C's Outboard sealed quart, 1957 (SS), $45-$95.

Sohio ATF A. Left: 1953 (SS). Right: 1962 (SS). One-quart cans, $25-$35.

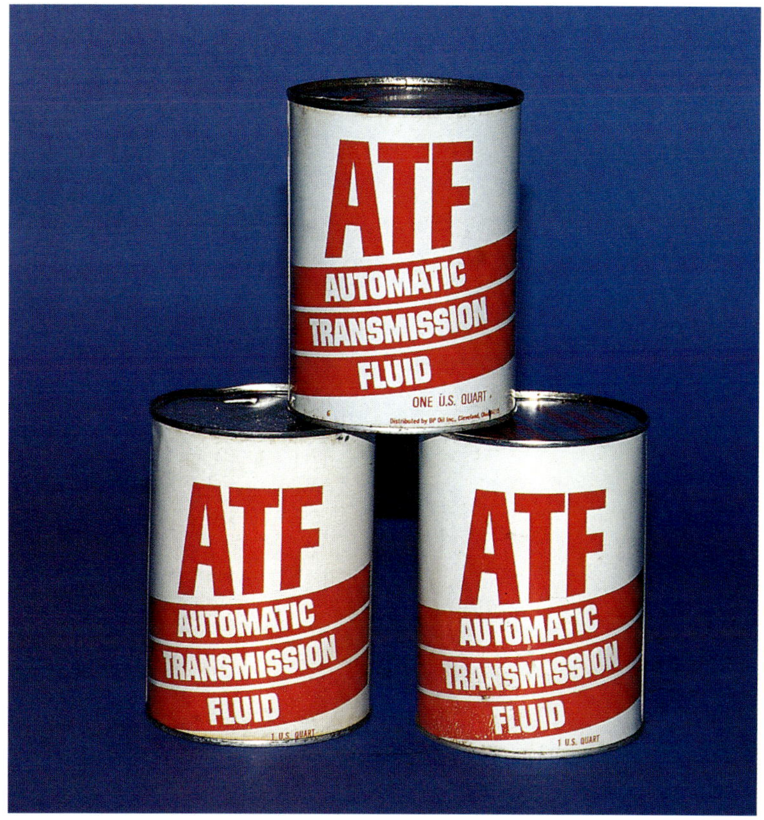

Sohio ATF. Top: 1969 (SS). Left: 1966 (SS). Right: 1964 (SS). One-quart cans, $25-$35.

Super Sohio Anti-Freeze. Left: 1935 (SS) one-quart can. Right: 1940 (SS) one-gallon can. $65-$125

Super Sohio Anti-Freeze. Top: 1959 (SS). Left: 1952 (SS). Right: 1956 (SS). One-quart cans, $25-$45.

Sohio HD S-3. Left: 1953 (SS). Right: 1962 (SS). One-quart cans, $75-$125.

Sohio HD. Left: 1955 (SS). Right: 1962 (SS). One-quart cans, $75-$125.

Sohio Octron. Left: 1962 (SS). Right: 1964 (SS). One-quart cans, $25-$40.

Left: Sohio Cetron, 1964 (SS), one-quart can, $25-$35. Right: Sohio CHD, 1966 (SS) one-quart can, $35-$45.

Left: Sohio Renown, 1935 (SS) two-gallon can. Right: Sohio Polarine, 1941-1967 (SS) two-gallon can. There are five variations of this can. $25-$100 each.

Sohio Products, 1946 (SS) five-gallon can. $75-$125

Sohio Outboard. Left: 1954 (SS) one-quart can. Right: 1957 (SS) one-quart can. $25-$35 each.

Sohio Household Oil, 1938 (SS) 4 oz. $75-$200

Sohio Household Oil. Left: 1939 (SS) 4 oz. Right: 1941 (SS) 4 oz. $50-$90

Fleetwing Oil, all 1949 (SS) one-quart cans. $45-$125

Sohio Lighter Fluid. 1953 (SS) 4 oz. $75-$150

Refiners Motor Oil, 1925 five-gallon can (SS). $50-$100

Fleetwing Piston Seal (SS), 1952 five-quart can. $45-$95

Fleetwing Oil cans, all 1963 (SS) one-quart cans. $25-$55

Fleetwing Super Anti-Freeze, 1941 (SS) one-quart can. $75-$125

Magic Motor Oil (Fleetwing). 1930s (SS) two-gallon can. $75-$175

Fleetwing two-gallon cans(SS), 1940-1960. $25-$45

Canfield Oil cans, all 1940-1950 (SS) one-quart cans. $25-$50

Canfield Premium. Left: 1950 (SS) one-quart can. Right: (SS) half-quart can. $25-$30 each.

Canfield Dependable Penn. 1950 (SS) one-quart cans, two variations. $25-$50

Canfield Premium, 1956 (SS) one-quart can. $125-$175

Canzol Lighter Fluid. 1930s (SS) 8-oz can. $75-$125

Wm Penn oil cans. All 1955-1965 (SS) one-quart cans. $25-$35

Standard Oil of Kentucky, 1962 (SS) 1-lb. can. This is one of the few cans with the S.O. KY logo as they sold Chevron motor oil and Esso and Mobil prior to 1962. Other known cans are a S-1 one-quart, household oil, and five-gallon product cans. $25-$45

State Motor Oil, 1935 (SS) two-gallon can. $30-$60

Sterling Oil Co., Oil City, PA

Sterling Motor Oil, Sterling Oil Co., Oil City, Pennsylvania. 1948 (SS) one-quart can. $40-$60

Sun Oil Co., Philadelphia, PA

Sunoco Motor Oil. Top: 1946 Folded seam (SS). Left: 1937 (SS). Right: 1936 Solder seam (SS). One-quart cans, $30-$50.

Sunoco Dynalube. Left: 1946 (SS). Right: 1951 split seam (SS). One-quart cans, $40-$75.

Sunoco Dynalube. Left: 1960 (CP). Right: 1968 (CP). One-quart cans, $25-$35.

Sunoco Dynalube. Left: 1951 Folded seam (SS). Right: 1952 Folded seam (SS). One-quart cans, $40-$75.

Sunoco Special/Dynalube, 1960 (SS) five-quart can. Also sold in one-quart cans. $40-$60

Sunoco Dynalube. Top: 1956 (SS). Left: 1952 Crimp seam (SS). Right: 1956 (SS). One-quart cans, $40-$75.

Sunoco Special. Left: 1956 (SS). Right: 1960 (CP). One-quart cans, rare, $30-$60.

Sunoco Special. Top: 1973 (CP). Left: 1971 (CP). Right: 1967 (CP). $25-$35 each.

Sunoco Custom, 1965 (CP) one-quart can. Rare, $50-$100.

Sun-Motor. Top: 1963 (CP). Left: 1957 (SS). Right: 1960 (CP). One-quart cans, $25-$50

Sunoco Mercury. Left: 1961 (SS). Right: 1963 (CP). One-quart cans. Both also sold in two-gallon cans. $25-$40

Sunlube. Top: 1973 (CP). Left: 1967 (CP). Right: 1969 (CP). One-quart cans, $25-$35.

Sunoco Diamond, 1971 (CP) one-quart can. $15-$25

Sunfleet. Top: 1973 (CP). Left: 1960 (SS). Right: 1965 (CP). One-quart cans, $25-$50.

Sunoco Motor Oil. Left: 1936 (SS) two-gallon can. Right: 1960 (SS) two-gallon can. $35-$125 each.

Sunoco Motor Oil. 1937 (SS) two-gallon can. $25-$40

Sunoco Transmatic. Top: 1967 (CP). Left: 1949 (SS). Right: 1960 (CP). One-quart cans, $25-$75.

Sunoco Motor Oil, 1956 (SS) two-gallon can. The 1946 can (not pictured) has the same design as the 1956. $25-$35

Sunoco Anti-freeze. Left: 1969 (SS). Right: 1973 (SS). One-quart cans, $25-$35.

Sunoco Anti-freeze, 1955 (SS) one-gallon can. Also sold in quart size. $30-$50

Sunoco Outboard. Left: 1949 (SS). Right: 1953 (SS). One-quart cans, $40-$60.

Sunoco Outboard. Left to right: 1956 (SS), 1958 (SS), 1960 (SS). One-quart cans. $35-$45

Tenneco Oil Co., Houston, TX

Tenneco Four Seasons, Tenneco Oil Co., Houston, Texas. 1961 (SS) one-quart can. $30-$50

Tenneco, All 1979 (CP) one-quart cans. $10-$20

Bay Petroleum Co., Houston, TX; Denver, CO

Bay Motor Oil, Bay Petroleum Co., Houston, Texas; Denver, Colorado. 1960 (SS) one-quart cans. $75-$150 (Tenneco)

Texaco Inc., New York, NY

Texaco Motor Oil. Left to right: 1934 (SS), 1936 (SS), 1941 (SS). One-quart cans, $25-$60.

Texaco Motor Oil. Left to right: 1947 (SS), 1953 (SS), 5/59 (SS). One-quart cans, $25-$50.

Texaco Motor Oil. Left: 5/62 (SS). Right 1/63 (SS). One-quart cans, $35-$45.

Texaco Havoline. Left to right: 1934 (SS), 1940 (SS), 1947 (SS), 1953 (SS), 9/58 (SS). One-quart cans, $20-$60.

Texaco Havoline 10W30, both 12/54 (SS) one-quart cans. Variation in top and gold finish. $30-$45

Texaco Havoline 10W30. Left: 9/58 (SS). Right: 5/59 (SS). One-quart cans, $30-$45.

Left: Texaco Havoline All Temperature, 5/62 (SS) one-quart can. Right: Texaco Havoline, 1/63 (SS) one-quart can. $20-$30 each.

Texaco Havoline 10W30. Left: 7/79 (SS). Right: 6/77 (CP). One-quart cans. $25-$35

Texaco Havoline All Temperature. Top: 7/69 (SS). Left: 8/67 (SS). Right: 10/68 (SS). One-quart cans, $20-$30.

Texaco D303. Left: 3/56 (SS). Right: 5/59 (SS). One-quart cans, $50-$100.

Texaco Havoline Super Premium. Left: 4/71 (SS). Right: 4/71 (CP). One-quart cans. $15-$25

Texaco Ursa S1, Left 6/65 (SS), Right: 10/61 (SS) one-quart cans. $30-$45 each

Left to right: Texaco Ursa, S-3-8/68 (SS), one-quart can. LA-3 2/72 (SS) one-quart can. ED-1/77 (SS), one-quart can. Superplus-7/79 (SS) one-quart can. $25-$35 each.

Texamatic Fluid. Left: 5/59 (SS). Right: 12/61 (SS). One-quart cans, $35-$45.

Top: Texaco Motor Detergent, 1953 (SS) one-quart can, $25-$35. Left: Texaco Ursa Tex, 3/72 (SS), one-quart can, $15-$30. Right: Texaco Ursa Tex, 7/78 (SS), one-quart can, $15-$30.

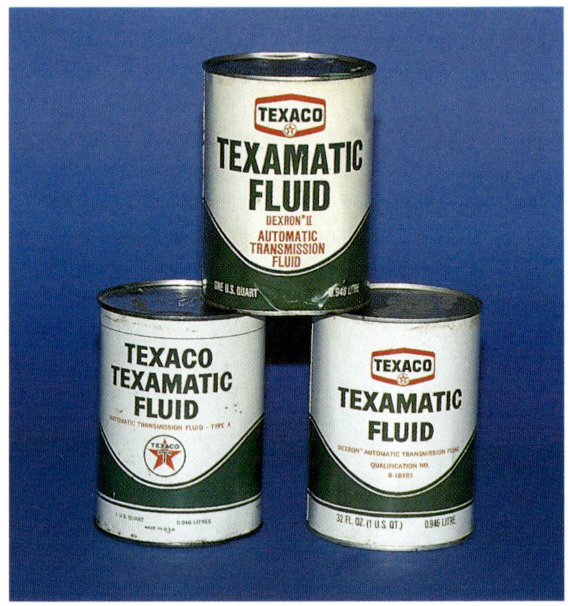

Texaco Texamatic. Top: 7/79 (SS). Left: 1/65 (SS). Right: 11/68 (SS). One-quart cans, $25-$35 each.

Texaco Valor. Left: 1934 (SS) two-gallon can. Right: 1946 (SS) two-gallon can. $45-$75

Texaco PT. Left: 1953 (SS). Right: 3/58 (SS). One-quart cans, $40-$75 each.

Texaco Anti-Freeze. Left: 4/62 (SS). Center: 3/66 (SS). Right: 4/72 (SS). One-quart cans, $30-$45 each.

Texaco Premium Aircraft Oil. 2/65 (SS) and 6/68 (SS) one-quart cans. $25-$35

Texaco Aircraft. Left: 1953 (SS). Right: 12/61 (SS). One-quart cans, $50-$75 each.

Texaco Marine Oil. Left: 1946 (SS). Right: 1955 (SS). One-quart cans, $200-$500 each.

Texaco Aircraft Oil. 11/64 (SS) and 5/68 (SS) one-quart cans. $25-$35

Texaco Outboard. Top: 5/59 (SS). Left: 1/70 (SS). Right: 1/73 (SS). One-quart cans, $35-$75 each.

Texaco Outboard, 5/66 (SS) one-quart can. $40-$60

Texaco Universal Gear Lubricant. 1953 (SS) one-gallon can. $50-$75

Texaco Outboard. Left to right: 9/54 (SS), 3/57 (SS), 1/61 (SS). $30-$75

Texaco Outboard Gear Oil. Left: 1949 (SS). Right: 1954 (SS). One-quart cans, $40-$70.

Leonard Oil Co., Alma, MI

Leonard Safe-T-Lube, Leonard Oil Co., Alma, Michigan. (CP) one-quart can. $40-$75

Leonard Outboard. Left: 1957 (SS). Right: 1964 (SS). One-quart cans, $40-$60.

Total Leonard, Alma, MI

Total, Total Leonard, Alma, Michigan. All 1979 (CP) one-quart cans. $15-$25

Union Carbide Consumer Products Division

Prestone Anti-Freeze, Union Carbide Consumer Products Division. Left: 1961 (SS) Right: 1965 (SS). $15-$30 each.

United Refining Co., Warren, PA (Keystone Stations)

Left: Powerlube Motor Oil, United Refining Co., Warren Pennsylvania (Keystone stations). 1970 (SS) one-quart can. $25-$35. Right: Emblem ATF, 1970 (SS) one-quart can. $25-$35

Union Oil Co. of California, Los Angeles, CA

Triton Motor Oil, 1934 (SS) one-quart can. $40-$60

76 Union Royal Triton 10W30. Left to right: 1956 (SS), 1959 (SS), 1961 (CP). One-quart cans, $25-$35.

76 Union Royal Triton. Left: 1950 (SS). Right: 1956 (SS). One-quart cans, $30-$40.

Royal Triton 10W30, 1956 (SS) one-quart can. $30-$40

76 Union Royal Triton, 1953 (SS) one-quart can. $30-$50

76 Union Royal Triton. Left to right: 1955 (SS), 1957 (SS), 1959 (AL), 1961 (CP). One-quart cans, $25-$35.

Left: 76 Union Super Royal Triton, 1960 (SS) one-quart can. Right: 76 Union Royal Triton, 1960 (SS) one-quart can. $35-$50 each.

Left to right: 76 Union Super Royal Triton, 1969 (CP) one-quart can; 76 Union Royal Triton, 1969 (CP) one-quart can, $25-$35 each. 76 Union Triton, 1969 (CP) one-quart can; 76 Union T5X, 1969 (CP) one-quart can; 76 Union ATF, 1969 (CP) one-quart can; $35-$50 each.

76 Union Triton. Left to right: 1953 (SS), 1955 (SS) 1957 (SS). One-quart cans. $25-35 each.

76 Union Triton. 1958 (SS) and 1959 (SS) one-quart cans. $30-$40

76 Union T5X. Left: 1954 (SS). Right: 1952 (SS). One-quart cans. $35-$45 each.

76 Union T5X Aero Oil, 1954 (SS) one-quart can. $40-$60

76 Union Anti-Freeze, 1954 (SS) one-quart can. $40-$60

76 Union ATF, 1955 (SS) one-quart can. $30-$60

76 Union Outboard, 1955 (SS) and 1947 (SS) one-quart cans. $35-$65

Union ATF, 1960 (AL) one-quart can. $30-$40

Resources

Buyer of oil cans, signs, gas station or any Sohio items:

W. Clark Miller
9944 Gardenside Drive
Waite Hill, OH 44094-6949
(440) 942-1079

Trade publications for collectors:

Petroleum Collectibles Monthly
P.O. Box 556
LaGrange, OH 44050-0556
(440) 355-6608 FAX (440) 355-4955

Check the Oil
P.O. Box 937
Powell, OH 43065-0937
(614) 848-5038

Sources for cans, signs, pumps etc.

Vic's Place
124 North 2nd
Gutherie, OK 73044

Ed Copelin
P.O. Box 181
Fairfield, ME 04937

Mr. Don Raleigh
42 Pine Hill Road
Stockton, NJ 08559

Bibliography

Rob Bender and Tammy Cannoy Bender, *An Unauthorized Guide to Mobil Collectibles*. Atglen, Pennsylvania, Schiffer Publishing Ltd., 1999

Scott Benjamin and Wayne Henderson, *Gas Globes: Amoco to Mobil and Affiliates*. Atglen, Pennsylvania, Schiffer Publishing Ltd., 1999

Scott Benjamin and Wayne Henderson, *Gas Globes: Penzoil to Union and Affiliates*. Atglen Pennsylvania, Schiffer Publishing Ltd., 1999

Scott Benjamin and Wayne Henderson, *Oil Company Signs: A Collector's Guide*. Osceola, Wisconson, Motor Books International, 1995

Scott Benjamin and Wayne Henderson, *Sinclair Collectibles*. Atglen, Pennsylvania, Schiffer Publishing Ltd., 1997

Scott Benjamin and Wayne Henderson, *Standard Oil Co.: The First 125 Years*. Osceola, Wisconsin, Motor Books International, 1996

Scott Benjamin and Wayne Henderson, *Texaco Collectors' Guide*. Osceola, Wisconsin, Motor Books International, 1997

Helms, Todd P. *The Conoco Collectors Bible*. Atglen, Pennsylvania, Schiffer Publishing Ltd., 1995

Wayne Henderson and Scott Benjamin, *Guide to Gasoline Logos*. Lagrange, Ohio, Marshall, North Carolina, PCM Publishing Co., 1997

Ed Love, Wayne Henderson, W. Clark Miller, *1990 Edition The Master List of United States Oil Containers*. Colorado Springs, Colorado, Gas House Publications, Division of Villa Publishing Syndicate, 1990

Love, Ed. *The American Oil Can Encyclopedia*. Colorado Springs, Colorado, Villa Publishing Syndicate, 1991

McIntyre, J. Sam, *The Esso Collectibles Handbook*. Atglen, Pennsylvania, Schiffer Publishing Ltd., 1998

Miller, W. Clark and Sabra Sonewald, *An Unauthorized Guide to Collecting Sohio*, Atglen, Pennsylvania, Schiffer Publishing Ltd., 2000

Moncrief, David L., *Got A Drop of Oil, An Introduction and Price Guide to Small Oilers*. Gas City, New Jersey, L-W Book Sales, 1998

Whitworth, Charles, *Gulf Oil Collectibles*. Atglen, Pennsylvania, Schiffer Publishing Ltd., 1998